* * *

Biomedical Research Ethics

by Mikyung Kim and Christopher B. Shook

ISBN 978-1-7344262-1-2 (Paperback Edition)

ISBN 978-1-7344262-0-5 (Digital Edition)

Table of Contents

1. Introduction 3

2. A Researcher's Role 5

3. SNU Internal Policies........ 11

4. Korean Government Laws........ 16

5. Universal Norms 22

6. Research Misconduct........ 26

7. Misconduct Types 33

8. Fabrication 36

9. Falsification 50

10. Plagiarism........ 57

11. Investigations and Whistleblowers........ 68

12. Authorship........ 83

13. Data Management........ 94

14. Publication 101

15. Peer Review........ 106

16. Reproducibility 110

17. Cognitive Bias 117

18. Conflict of Interest........ 119

19. Human Subjects and Informed Consent 128

20. IRB........ 136

21. Animal Experiments........ 139

22. Collaborative Research 144

23. Mentoring 149

24. Conclusion........ 151

1. Introduction

This course textbook covers clinical research ethics, emphasizing the real world, practical applications of the principles, examples, and rules that are included. The purpose of these materials serve not only to inform students of the prevailing and mandatory ethical norms and expectations of researchers, but also to foster an environment of ethical research within the Seoul National University academic community.

Consequently, the concepts and ideas here are presented from multiple viewpoints within the university community. As members of an internationally competitive research university, it is important for us to maintain the highest standards and quality in our science and research activities. To this end, international, including American, clinical research practices are incorporated as a complementary perspective to illustrate different means and methods of guiding researchers in ethically ambiguous situations. Specific notes about Korean cultural norms and research ethics are contrasted against competing global interpretations.

Finally, when reading this work, keep in mind that clinical research is a continuously evolving area. Frequent discussion with colleagues, university staff, and other resources can facilitate a better understanding of research ethics for individual situations. Many of the situations in these chapters anticipate the types of ambiguous events that researchers may encounter in the overall development of their careers. Ultimately, ethical research behaviors should be voluntary, in order to avoid the manipulation of ethical guidelines to cover up unethical actions. For some students, this course textbook may be the first time they are introduced to these research ethics concepts. If some of these situations seem new or surprising, do not worry. Rather than simply a list of rules and concepts to memorize, hypotheticals are offered to reveal a more practical view of ethics, as well as the rationale and reasoning for such behaviors. The purpose of this course is to prepare you and provide you with the necessary tools to confront and adapt to challenging ethical problems when they occur in the present or in the future.

2. A Researcher's Role

Whether someone is working in a lab as a researcher or writing an academic article, research ethics are a pervasive issue in the life of scientific and clinical knowledge workers in the modern economy. Researchers serve a public role in searching for cures for diseases, verifying experiments, testing new theories, and many other societal benefits. A single researcher working in isolation has the potential to revolutionize accepted scientific wisdom, but can also destroy and bring harm to an entire community. A team of many researchers can also offer much potential for answering fundamental questions about our world, with the potential to affect the credibility of entire branches of science. Inaccurate or misleading research can persuade other researchers to waste time and money on pointless or ineffective areas to replicate and advance results.

The general public often has very different ideas about what research involves. The people who benefit from research may only indirectly support the difficult and important work that results from years of effort. Many times, scientific research may continue even when it has only a slight chance of transforming our understanding. Why do researchers engage in these arduous tasks? Various incentives, ranging from monetary compensation, professional fame, and intellectual curiosity are but a few of these potential rewards. In fact, the role of researchers in pursuing these prizes often directly opposes the motivations and means for the ethical conduct of research. This tension, between ethics and financial goals, often confronts researchers during the course of their careers. If you have not yet faced this type of ethical dilemma, do not worry—someday, you will likely have to resolve an ambiguous or complex ethical situation where the proper course of action and the desirable course of action diverge.

In a way, for members of the Seoul National University academic community, we have an additional role as researchers. As a leading public institution that enjoys the goodwill and respect of this nation, this university imposes on researchers an additional role: that of role models. The behaviors and conduct of our academic community can influence other universities and researchers in a uniquely prominent manner. Therefore, it is crucial that we maintain especially high standards of ethical research both because of our role as researchers and as leaders.

When dealing with ethics, it can be difficult to avoid mentioning the concept of integrity. Integrity, and for our purposes, scientific integrity, is an important idea for researchers. Integrity refers to the ability and performance of a person in a way that reflect honest, ethical, and morally sound principles worthy of being emulated by others. In the absence of clear rules and guidelines for every single scenario, the scientific integrity of a researcher will determine how their ethical research reflects these ideals.

In contrast, when a researcher acts without integrity, many bad results can occur. In serious cases, ethical lapses by researchers can result in accusations of misconduct. One of the types of misconduct is research misconduct, and it is a potentially dangerous issue for the career of any researcher. Research misconduct refers to unethical behavior by a researcher that is not simply a minor mistake or accident, but an especially harmful violation of academic integrity.

It may or may not be immediately apparent that misconduct has occurred. Many institutions have official procedures for dealing with allegations of research misconduct.[1] However, outside of official policies and offices, a researcher should

[1] 연구진실성위원회. Retrieved from http://snuethics.snu.ac.kr/committee

have the individual capability to understand the social responsibility beneath all research. Science is public good, which has extrinsic benefits for society as a whole. Researchers engaged in pushing the frontiers of scientific innovation and the search for knowledge are supported by public faith and confidence. When research misconduct occurs, it casts into doubt an entire scientific research community and has effects that can extend far beyond the individual, team, or university at fault. This is especially true for clinical research.

Historically, freedom of science and academic freedom have never been absolute. Think about Galileo, who lived his final nine years under house arrest, because his celestial ideas were deemed heretical. At the time, scientific research was not free from theological doctrine.

Even now, some restrictions on the peer review process and on the flow of scientific information can be justified by being for biosafety, biosecurity, and biodefense, in addition to commercial interests in intellectual property rights. Particularly, biological information and technologies have suffered from the dual-use problem of research. Recently, two examples are the publication in *Nature* and *Science* of the reconstruction of the 1918 Spanish influenza pandemic virus in 2005; and of the genetically altered bird flu virus transmissible in mammals in 2012. The reason why biological science should be socially responsible is related to the fact that biological science is employing relatively basic, inexpensive, commonly used technologies compared to other technologies, such as atomic nuclear science.

However, it is unclear how much the restriction on the information flow would be effective to maintain public health and national security, when considering how much effectively future bioterrorists can self-educate and replicate the results themselves by reading the publications. Moreover, there are several examples that indicate open science and dissemination of information may promote public health and can be quite effective in finding a solution to public health problems such as SARS.

Another dimension is whether we are allowed to use all technologies available for all purposes. Set aside the safety and efficacy issues, many new technologies have been debated ethically. The examples are contraception, DNA recombination, human cloning, genomic editing, etc.

When misconduct occurs, it can result in suspicion and isolation between scientists. A single researcher engaged in misconduct can negatively affect the communication between other scientists far removed from the social networks and proximity of the bad actor. Because of the role of research in stimulating intellectual discourse and debate among scholars, misconduct can stifle the necessary free exchange of ideas among the scientific community.

Additionally, the highly visible nature of clinical research (e.g., clinical trials) endows researchers with a responsibility to the public. Because medicine has the potential to harm people severely as a result of new advances, the field has a high standard for researchers generating evidence. A complete and truthful evaluation of risks and benefits is necessary for new technologies and procedures to gain widespread adoption and approval. Government agencies, private companies, and other groups turn to scientific publications and studies for assistance in policymaking, product strategy, and the allocation of financing or funding. Prominent research scientists are often called to testify and explain their work for government commissions, judicial bodies, and media organizations. A single instance of misconduct can have far-reaching consequences.

Infamous instances of unethical research in Korea have brought shame to our country in the past. By only engaging in ethical research, you represent yourself, your profession, your team, your institution, and your national colleagues with the appropriate honor and responsibility that befit a leading international research university such as SNU.

3. SNU Internal Policies

Seoul National University's governance is overseen by a Board of Trustees, the President, and Auditors. The President directs the general affairs of the university, and three Vice-Presidents assist the President with these duties. SNU dedicates one of these three Vice-Presidents, the Vice-President for Research, solely to overseeing research.

Reporting to the President's branch of institutional governance are several Deliberative Bodies. One of these Deliberative Bodies is the Research Integrity Committee, and it is established by university regulations.

There are three other independent research ethics groups reporting to the President: the

Institutional Review Board, the Institutional Animal Care and Use Committee, and the Institutional Biosafety Committee.

Currently comprised of 15 members, the Institutional Biosafety Committee is tasked with maintaining biosafety by evaluating potential safety risks and scientific validity of research on genetically engineered organisms and pathogens. The Seoul National University Institutional Animal Care and Use Committee (SNUIACUC) manages the ethical handling of laboratory animals and animal experiment facilities to ensure the appropriateness of production, introduction, management, experimentation, usage, and disposal of laboratory animals.[2] The IACUC may also be known as the Animal Experiment Ethics Committee.[3] The Committee on Research Integrity (CRI) initiates various measures to prevent research cheating in research before it can occur, as well as to investigate and process claims of cheating in research when actual cases are alleged. The CRI is comprised of nine members appointed by the university president. Finally, the Bio Ethics Deliberation Committee handles the biosciences and ethical research plans, including the issues of consent from human subjects, DNA information, and tissue storage.[4] An independent research

---

[2] 공지사항. Retrieved from https://iacuc.snu.ac.kr/index.htm

[3] 연구윤리팀. Retrieved from http://snuethics.snu.ac.kr/

committee, the Future Research Committee, reports directly to the Vice-President for Research.

SNU also has several Administrative Organizations, such as the Office of Academic Affairs or the Office of Admissions. Similarly, the Office of Research Affairs is an Administrative Organization that monitors research ethics and other research issues. The Office of Research Affairs (ORA) reports to the Vice-President for Research.[5]

ORA administers all research issues for SNU. It has three main sections: the Division of Research Policy, the Division of Research Support, and the Research Ethics Team.[6] The Research Operations Committee also is a part of ORA.[7] The Research Ethics Team also manages research ethics assessments, training, and investigations of ethical violations.

Finally, the SNU R&D Foundation helps with

---

[4] 최신연구동향 - 서울대학교 연구처·산학협력단. Retrieved from http://snurnd.snu.ac.kr/web/eng/einf_01

[5] Seoul National University. Administration. Retrieved from http://www.useoul.edu/research/administration

[6] Seoul National University. Organizations. Retrieved from http://www.useoul.edu/organization/administrative-organizations

[7] 최신연구동향 - 서울대학교 연구처·산학협력단. Retrieved from http://snurnd.snu.ac.kr/web/eng/about_04

research support and private industry.[8] The R&D Foundation shares resources with ORA.

All of these units within the university work together to promote organizational goals of ethical research principles. Many of these ideals have been formalized in internal written codes.[9] They may also derive from the Sciences Promotion Act and the National R&D Innovation Act, among other sources.

Seoul National University Research Ethics Guidelines Revised on July 16, 2010. Preamble Seoul National University (hereinafter referred to as "SNU") is not only an institution of higher education producing outstanding individuals but also a leading Korean research institute boasting of research scholars such as faculty, researchers, and graduate students. SNU's researchers, through academic advancement, have great responsibilities to set examples of contributing to both Korean society's development and the mutual prosperity of mankind.

[8] Brochure. Retrieved from http://www.useoul.edu/upload/research/brochure_eng.pdf

[9] 최신연구동향 - 서울대학교 연구처·산학협력단. Retrieved from http://snurnd.snu.ac.kr/web/eng/einf_01

Seoul National University Research Ethics Guidelines

Revised on July 16, 2010.

Chapter 8 Education on Research Ethics

Article 31

【Duties of Researchers】

① All researchers shall have thorough knowledge of the contents of the Research Ethics Guidelines and shall comply with them.

② All researchers shall actively participate in the research ethics education provided by Seoul National University, and a principal investigator shall lead researchers and students under his/her control so that they participate in the education.

Discussion:

1. Read the Preamble to SNU guidelines on ethical research. What does the organizational structure of SNU's research ethics system suggest about the university's commitment to supporting ethical research?

2. Note the various different committees that SNU uses to monitor ethical research, including some committees that function independently outside of any single college. Why do you think that independence is important for an oversight committee?

4. Korean Government Laws

In addition to professional and university codes of conduct, government laws and regulations can describe and enforce ethical research. This includes legal regimes on the use of government funding, the protection of animal and human subjects, and responsible safety practices. Government laws monitor behavior and provide unambiguous, prohibitions and requirements. This includes legal regimes and regulations.

Recall that SNU's Committee on Research Integrity deals with research ethics and integrity. This committee was established to prevent research misconduct before it occurs and to perform pre-investigations, investigations, and investigation results reporting for the university when such cases occur.[10] Duties extend to nine members, including Office of Research Affairs leadership, Office of Educational Affairs leadership, among others.

This Committee on Research Integrity was based on SNU regulations, Ministry of Education guidelines, and the Science Promotion Act [학술진흥법] (Law No. 10877, enforcement on July 21, 2011; as of January 11, 2019, Law No. 14163).

---

[10] 최신연구동향 - 서울대학교 연구처·산학협력단. Retrieved from http://snurnd.snu.ac.kr/web/eng/einf_01

These include the Directive for the Upholding Research Ethics from the Ministry of Education and the Enforcement Regulation of the National R&D Innovation Act from the Ministry of Science and ICT. Other ethical research committees also have their basis in government laws. For laboratory and animal issues, the SNUIACUC was based on the Animal Protection Act [동물보호법] (Law No. 10310, enforcement on November 26, 2010; as of January 11, 2019, Law No. 15502, amended enforcement on March 20, 2018) and the Laboratory Animal Act [실험동물에관한법률] (Law No. 9932, enforcement on March 19, 2010; as of January 11, 2019, Law No. 11987, partially amended enforcement on July 20, 2013). These two acts require protection for animals involved in experiments, which the SNUIACUC oversees.

The Biosafety Committee, meanwhile, is based on the Trans-boundary Movement of Living Modified Organisms Act [유전자변형생물체의국가간이동등에관한법률] (Law No. 9932; 15181) and the Biotechnology Support Act [생명공학육성법] (Law No. 10872; 14839). The Bioethics Committee derives its authority from the Bioethics and Safety Act [생명윤리및안전에관한법률] (Law No. 10605; 15188).

Other relevant laws and sources include the National R&D Innovation Act and presidential decrees.

Animal Protection Act

Article 23 (Principles for Animal Testing)

(1) Animal testing shall be conducted in consideration of the enhancement of welfare of humankind and the dignity of lives of animals.

(2) When it is intended to conduct animal testing, an alternative method shall be taken into consideration prior to such testing.

(3) Animal testing shall be conducted by a person who has knowledge and experience in ethical management and scientific use of animals used for testing (hereinafter referred to as "laboratory animals"), and the minimum number of animals shall be used.

(4) Animals less sensitive to pain shall be used for testing that causes pain to laboratory animals, and measures appropriate for relieving such pain by a veterinary method, such as the use of an analgesic, sedative, or anesthetic, shall be taken.

(5) A person who completes animal testing shall examine the animal without delay, and shall dispose of it as quickly as possible by a method that does not cause any pain, if the result of the testing shows that the animal cannot recover or is likely to live with a persistent pain.

(6) Except as otherwise provided for in paragraphs (1) through (5), matters necessary for the principles of animal testing shall be determined and publicly notified by the Minister of Agriculture, Food and Rural Affairs. <Amended by Act No. 11690, Mar. 23, 2013>

Bioethics and Safety Act

Article 3 (Basic Principles)

(1) No activity regulated under this Act shall be conducted in any manner that violates the dignity and values of a human being, and priority shall be given to human rights and welfare of each human subject of research or donor.

(2) Self-determination of each human subject of research or donor shall be respected, and the voluntary consent of a human subject of research or donor shall be supported by adequate information.

(3) Privacy of each human subject of research or donor shall be protected, and personal information likely to lead to the invasion of privacy shall be protected as confidential information, except where the relevant party consents to disclosure or an Act expressly permits disclosure.

(4) Full consideration shall be given to the safety of each human subject of research or donor, and risks shall be minimized.

(5) An individual or group in vulnerable conditions shall be specially protected.

(6) International cooperation shall be promoted as necessary to ensure bioethics and biosafety, and a person conducting such activities shall endeavor to adopt universal norms.

Bioethics and Safety Act

Article 20 (Prohibition of Human Cloning)

(1) No person shall implant a somatic-cell

cloning embryo or parthenogenic embryo (hereinafter referred to as "somatic-cell cloning embryo") into a human or animal womb or keep such embryo implanted or bear a child therefrom.

(2) No person shall solicit another person to conduct an activity referred to in paragraph (1) or act as a broker for such activity.

Article 21 (Prohibition on Implantation between Different Species)

(1) No person shall implant a human embryo into an animal womb or implant an animal embryo into a human womb.

(2) No person shall engage in the following activities:

1. Fertilizing a human ovum with an animal spermatozoon or an animal ovum with a human spermatozoon: Provided, That medical tests for examining the activity of human spermatozoa shall be excluded herefrom;

2. Implanting an animal somatic nucleus into a human ovum with its nucleus removed or a human somatic nucleus into an animal ovum with its nucleus removed;

3. Fusing a human embryo with an animal embryo;

4. Fusing human embryos with different genetic information.

(3) No person shall implant a thing produced from a procedure referred to in any subparagraph of paragraph (2) into a human or animal womb.

Discussion:

1. Both the Bioethics and Safety Act and the Animal Protection Act make it clear that the dignity of humans and animals should be protected. Why should a researcher be concerned about dignity?

5. Universal Norms

There are some ideas about research ethics that are so basic, so fundamental that they are nearly universal in their applicability to research scholars. This section describes selected principles of international research, with the understanding that research in Korea is frequently subject to analysis and criticism from renowned scholars abroad. Later sections describe in depth relevant global perspectives of specific research ethics topics.

Honesty refers to the ability of a researcher

to accurate engage in truth and to openly state facts and opinions. When a researcher is questioned about the veracity of their work, a researcher must disclose important details even if it is easier to remain silent.

Fairness refers to how a researcher negotiates the landscape of various people involved in researcher. Because research can often be expensive and time consuming, resources may be scarce. Laboratories, equipment, and staff may be limited and only offer certain opportunities for select individuals. Fairness ensures that resources are allocated in ways that prioritize resources and research over self-interests.

Professionalism, an increasingly important tool for ethical researchers, deals with researchers and their internal codes of conduct specific to their profession. The ethical conduct of research constantly evolves in response to challenges and confrontations by unethical or ignorant practices. Professionalism is key to resolving the clash of codes that can occur when outdated or ignorant research practices come in contact with updated or rigorous ethics.

Objectivity and maintaining a level, balanced outlook on situations is especially necessary in situations involving suspected misconduct. The neutral perspective towards the performance of research ensures that even in cases where potentially inappropriate research behaviors have been observed, just and standardized processes evaluate those events.

Trust underpins the foundation of research. When journals publish findings, they rely on the trust of the academic community to accept research as being truthful and accurate. Without trust in research, the ability of ideas to disseminate throughout universities and countries would be harmed. Trust in ethical research allows for mutual collaborations and globalized breakthroughs.

Respect, often respect for others, is norm that is frequently misunderstood. The special circumstances that researchers face can result in very rigid hierarchies and situations involving authority or groupthink. Respect for others, including others' ideas, means that the critique and analysis of potential research conclusions never loses sight of the fact that colleagues are shared participants in the search for knowledge and understanding through research. Often, standards, customs and accepted practices may differ from lab to lab or between specific scientific disciplines, and respect furthers this goal of working smoothly with others.[11]

At the same time, there potentially could be various ethical issues specific to Korean society. Confucian legalist principles that have fundamental influence in many areas of life can result in

---

[11] *On being a scientist: A guide to responsible conduct in research*. (2009). Washington, D.C.: National Academies Press. doi:10.17226/12192

different ways of thinking from other places. This is also true for scientists conducting research, who may grapple with collectivism, interdependence, and comprehensiveness in ways that more individualistic research environments in other places do not.[12] Whereas Western university culture may value the autonomy of the independent researcher, East Asian society may value the honor of groups. This extends to areas such as whistleblowing, joint research, and conflicts of interest, areas that can be difficult to import from Western research ethics.[13] To mitigate these gaps, emphasizing basic concepts such as fairness and respect for human rights can reinforce human subject protections, the principles of peer review, mentoring, and investigations of misconduct.[14]

---

[12] Nho, H. (2016). Research ethics education in Korea for overcoming culture and value system differences. *Journal of Open Innovation: Technology, Market, and Complexity, 2*(1). doi:10.1186/s40852-016-0030-3

[13] Nho, H. (2016). Research ethics education in Korea for overcoming culture and value system differences. *Journal of Open Innovation: Technology, Market, and Complexity, 2*(1). doi:10.1186/s40852-016-0030-3

[14] Nho, H. (2016). Research ethics education in Korea for overcoming culture and value system differences. *Journal of Open Innovation: Technology, Market, and Complexity, 2*(1). doi:10.1186/s40852-016-0030-3

6. Research Misconduct

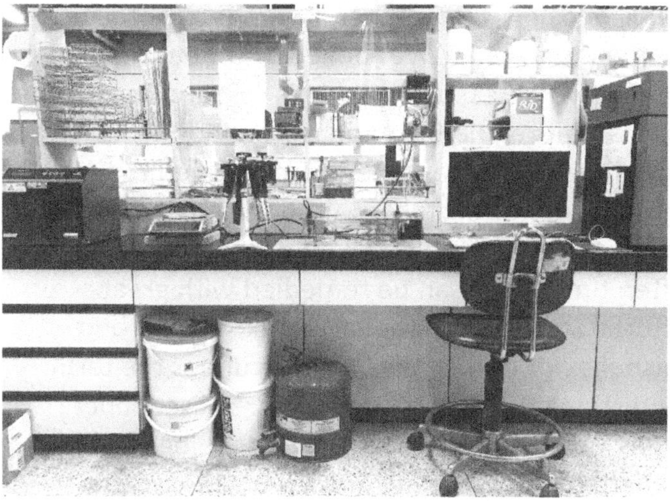

Sometimes researchers make mistakes. Sometimes researchers misbehave. Sometimes researchers fail to fulfill their ethical responsibilities.

Are these examples of research misconduct? Not necessarily. Research misconduct is an especially severe form of unethical research behavior. Misconduct is usually defined as behaviors that are not simply a mere mistake, an honest error, or a risk. Misconduct rises to a higher level of ethical breach with longer lasting effects. This can mean loss of reputation, effects to the community in aggregate, or damaged academic relationships.

Misbehavior, however, is distinguishable both for its effects and for its nature. Misbehavior may result from intentional acts or gross negligence by a researcher, perhaps a protocol or step that has been overlooked or unreasonably disregarded. Because academic research is in many ways a self-regulating endeavor, minor mistakes with little fallout can often be presumed to be misbehavior or mistakes. Although mistakes can harm trust in others and data or methodology, often times they can be remedied with great efforts. Some international sources may group misbehavior or mistake together under the term "questionable research practices." Misconduct many times requires that the underlying intent be to defraud or deceive.[15] However, this factor can often be difficult to prove to confirm misconduct.[16]

SNU's research guidelines give an exhaustive list of potential misbehaviors. These range from exaggeration, incorrect citations, honorary authorship, or duplicate publications, to even aiding someone in committing a more serious breach of misconduct. These topics are explored in later sections. Honest mistakes can happen in every field of study, and university guidelines note that the inquiry into misbehavior is field-specific. Because clinical and some scientific research has

---

[15] Shamoo, A. E., & Resnik, D. B. (2015). *Responsible conduct of research*. New York, NY: Oxford University Press.

[16] Stewart, C. N. (2011). *Research ethics for scientists: A companion for students*. Oxford: Wiley-Blackwell.

relatively higher ethical concerns than other branches of research, it is important for researchers in these fields to be aware of the differences between research ethics norms when conducting interdisciplinary research.

It would be easy to devote the entirety of this book to writing solely about scientific misconduct, specifically research misconduct. However, the goal here is not to train young researchers to fear ethics, but rather approach research ethics from a knowledgeable standpoint. The section herein describes specific instances of misconduct from around the world that aim to educate and elucidate those values and actions that make up research ethics. Research misconduct is all too commonplace in the 21st century, ranging from high profile journals to minor grant submissions. In general, the pervasive nature of falsified, fabricated data in academic publications can lead to retractions and much more severe consequences. To this end, education is power. Young researchers can substantially benefit from education on best practices to avoid unethical pitfalls in research, in conjunction with typical material on the negative aspects of unethical conduct.[17]

---

[17] Stewart, C. N. (2011). *Research ethics for scientists: A companion for students*. Oxford: Wiley-Blackwell.

Seoul National University Research Ethics Guidelines Revised on July 16, 2010. Chapter 2 Integrity in Research Presentation Article 13 【Judgment of Research Misconduct and Research Misbehavior】 The committing of an act falling under research misconduct and research misbehavior respectively under Article 11 and Article 12 and the degree of violation shall be judged considering the characteristics of a major field of study and the opinions of the relevant academic field.
Seoul National University Research Ethics Guidelines Revised on July 16, 2010. Chapter 2 Integrity in Research Presentation Article 12 【Research Misbehavior】 ① A researcher who has committed, not through serious negligence, an act falling under any of the subparagraphs of Article 22 in the entire course of research such as proposal/performance/report/presentation, etc., of a research shall be held liable for research misbehavior. ② A researcher who has committed,

intentionally or through gross negligence, an act falling under any of the following subparagraphs shall be responsible for such research misbehavior:

1. Act of deriving an untrue research result by interpreting the research materials in an exaggerated, reduced, or distorted manner in violation of Subparagraph 3, Article 6;

2. Act of using research idea, research data, and sentences by another person without correct indication of its source or citation in violation of Subparagraph 3, Article 7 but which does not fall under Subparagraph 3, Article 11;

3. Act of heavily relying on other person's idea, research data, and sentences to the extent that may affect the originality of his/her own research in violation of Subparagraph 4, Article 7 (regardless of whether or not indication of its source or citation has been made);

4. Act of using his/her own research idea, research data, and sentences that have already been published as if they were presented for the First time in such research in violation of Paragraph (1), Article 8;

5. Act of relying upon his/her own research idea, research data, and sentences that have already been published to the extent that affects the originality of such research in violation of Paragraph (2), Article 8 (regardless of whether or not indication of its source or citation has been made);

6. Act of publishing all or part of his/her own theses or writings that have already been published in the same language or other languages without correct indication of its source or citation

in violation of Article 9; 7. Act of including a researcher who has not made a direct contribution to research in the coauthors or intentionally excluding a researcher who has made a direct contribution to a research from the co-authors in violation of Paragraph (2) or (5), Article 10; or 8. Act of conniving, aiding, or concealing research misconducts specified in Article 11.
Ministry of Education Directive for the Upholding of Research Ethics July 17, 2018, Partial Amendment Chapter 3 Research Misconduct Article 12 (Scope of Research Misconduct) ① Research misconduct includes the following items committed in any or all stages of proposal, conduct, result presentation, and publication of research. 1. Fabrication: making up nonexistent research raw data, research data or results 2. Falsification: distorting research content or results by artificially manipulating materials, facilities and procedures, changing or deleting research raw data, research data or results 3. Plagiarism: stealing or copying others' original idea or research work without proper citation or acknowledgement and making people recognize that it is as one's own a. Using all or a part of others' research work or its content without properly citing the sources b. Modifying and using the words and the sentence structures of others' writings without

properly citing the sources c. Using others' original ideas without properly citing the sources d. Translating and using others' research work without properly citing the sources 4. Invalid authorship: Not giving authorship to those who contribute to research activities or results without legitimate reasons, and attributing authorship to those who did not or less contribute to research activities or results a. Attributing authorship to those who did not contribute to research contents or results

b. Not giving authorship to those who have contributed to research contents or results c. Publishing a student's research paper or degree thesis under his/her professor's name only

5. Unjust Duplicate Publication: publishing again his/her own previous research work itself or its modification substantially similar to the original version and gaining unjust profits such as more research funds or additional academic assessment scores

6. Interruption of research misconduct investigation: Interrupting the investigation process of one'sown or others' research intentionally or harming the complainants

7. Behavior beyond commonly tolerable scope of each research field

② In addition to Paragraph 1, universities and institutes may include other behaviors in the scope of research misconduct when they think those behaviors should be investigated or prevented

7. Misconduct Types

The three main types of misconduct are plagiarism, fabrication, and falsification. Fabrication refers to making up data or results, either when recording or reporting. Falsification occurs when manipulation distorts research materials, equipment, or methods, as well as when data or results are changed or left out to distort the overall research impression. Plagiarism occurs when the words, ideas, ideas, or academic content of another are taken without giving credit.

Although there are other types of egregious, unethical behavior that could be called misconduct, these three categories are the most common and they are commonly referred to via the acronym "FFP."[18]

Article 12 of the Directive for the Upholding of Research Ethics from the Ministry of Education also mentions these areas of misconduct as well as others. In some cases, authorship issues or interfering with an investigation into research may also qualify as forms of misconduct.

Researchers have developed these standards regarding misconduct because of centuries' worth of scientific progress. Until recently, they were set forth not as explicit rules,

---

[18] ORI. Definition of Research Misconduct. Retrieved from https://ori.hhs.gov/definition-misconduct

but as the professional guidelines on the acceptable and permissible ways of conducting research. Researchers who violate these standards can face a number of severe consequences, including retractions of publications, revocations of doctoral degrees, loss of tenure, employment termination, criminal penalties, and more. In a notable case, an employee with a famous Japanese research firm had a PhD revoked after questions arose about unclear and suspicious issues with the data on reprogramming adult cells to function as stem cells.[19] The researcher's collaborator on the research later committed suicide in the laboratory, in part due to shame and the consequences of the media coverage.[20]

It is clear, then, that those who would expose themselves to misconduct accusations may lose face, prestige, and the respect of their peers and potentially ruin their careers and the lives of others.

Seoul National University Research Ethics Guidelines Revised on July 16, 2010.

[19] Hermes. (2016, July 21). Scandals in the scientific community. Retrieved from https://www.straitstimes.com/singapore/scandals-in-the-scientific-community

[20] Spitzer, K. (2014, August 08). Japan: Stem-Cell Science Scandal Triggers Suicide, Soul-Searching. Retrieved from http://time.com/3091584/japan-yoshiki-sasai-stem-cells-suicide-haruko-obokata/

Chapter 2 Integrity in Research Presentation

Article 11

【Research Misconduct】

A researcher who has committed an act falling under any of the following subparagraphs in the entire course of research intentionally or through gross negligence (it refers to the case of significantly violating the ordinary duties of care as a researcher) such as proposal/performance/report/presentation, etc., of research shall be held liable for research misconduct:

1. Act of fabricating, recording, or reporting research data or research material in violation of Paragraph (1), Article 6;

2. Act of manipulating a research material by arbitrarily changing research data, adding something thereto, or leaving it out in violation of Paragraph (2), Article 6; or

3. Act of using research outcome by any other person as if it were his/her own research outcome in violation of Paragraphs (2) or (3), Article 7.

Discussion:

1. This section and the preceding section describe two types of unethical research, misbehavior and misconduct. What are the major differences between them?

8. Fabrication

As mentioned previously, fabrication is one of the three main types of misconduct in research. It is also often viewed as the most severe among the three, being comparatively more harmful than falsification or plagiarism. Fabrication commonly involves researchers who make up or create data for experiments that have not actually been performed or results that have not been achieved. However, fabricated data is not the only type of fabrication. Some academics have come under fire for making up citations or even co-authors. These too can be potential instances of misconduct, though more minor in nature than data fabrication.

Fabrication can also be a mark of an especially unethical researcher, and a single instance of data fabrication can haunt the career of those accused. Hans Schöler of the Max Planck Institute for Molecular Biomedicine in Münster, Germany notes, "If you fabricated data once, how would one know that you will not do it again?"[21]

Therefore, young researchers must be especially aware of how to avoid data fabrication. Conducting research can be an especially demanding process, involving numerous people

---

[21] Cyranoski, D. (2014). Cloning comeback. *Nature, 505*(7484), 468-471. https://www.nature.com/news/cloning-comeback-1.14504#/timeline

and stakeholders. It is arguable that unethical research via data fabrication is a way of cheating to achieve an unfair advantage in competition with other researchers by avoiding having to undergo thorough and potentially hurried experimentation.[22] Not only does a researcher have to obtain funding, and recruit skilled employees, they also have to ensure that the experiments result in positive outcomes that show a hypothesis can be proven and supported by empirical evidence. This is because data that results in a rejection of the hypothesis typically fails to find a welcoming publication outlet for this type of result. At the same time, however, this negative result and rejection of a hypothesis could assist other researchers in eliminating a potential line of research. Given that clinical research is the search for knowledge, proving a negative in research is seen as not adding to the overall body of knowledge in a new way. Although this notion is debatable, especially given the push for studies that are capable of reproduction or reproducibility, the fact is that publication bias can make it difficult to publish studies that do not result in positive findings and innovations. Whereas fabricating data is much easier than having to actually conduct a rigorous study that may not result in positive findings, the benefits of being the first to publish on a certain new topic can reap great benefits. As such, scientific research can foster unethical behavior in the race for academic awards.[23]

---

[22] Stewart, C. N. (2011). *Research ethics for scientists: A companion for students*. Oxford: Wiley-Blackwell.

This perverse incentive was possibly an explanatory factor in the well-known case of a former professor at a famous Korean university. In the early 1990s, South Korea announced plans to promote a local biotechnology industry.[24] In 1994, the South Korean government and other businesses allocated $18 billion in funding for research and development in biotechnology, with the hope that by 2010 South Korea would be one of the leading biotechnology nations.[25] Seoul, with its local favor of *bbali bbali* and desire for doing things quickly rather than methodically, was the incubator for this professor's rapid rise to the top of global science. At the same time, it was also a factor in the misconduct coverage that plagues Korean scientific research.

At the time of discovery of misconduct, this professor had been a national hero for what the world believed to be research involving stem cell lines derived from human embryos that were cloned. This professor's portrait graced public buses and postage stamps, while posters described the way his apparent historic discovery symbolized his success and the hope for worldwide fame and

---

[23] Stewart, C. N. (2011). *Research ethics for scientists: A companion for students*. Oxford: Wiley-Blackwell.
[24] Gottweis, H., & Triendl, R. (2006). South Korean policy failure and the Hwang debacle. *Nature Biotechnology, 24*(2), 141-143. doi:10.1038/nbt0206-141
[25] Gottweis, H., & Triendl, R. (2006). South Korean policy failure and the Hwang debacle. *Nature Biotechnology, 24*(2), 141-143. doi:10.1038/nbt0206-141

prominence. Koreans viewed this professor as a vessel for global stem cell research opportunities, and they believed that helping this professor's research was a way to help the Korean people and country as well.[26] This was another unacknowledged source of pressure for this professor to fabricate his data: his fellow citizens had pinned their hopes and dreams for Korean biotechnology on his research. In fact, this professor embraced this sentiment and the appeals to Korean national identity that helped propel him to fame: shortly after cloning his first cow, this professor promised Korea that he would attempt to further Korean spirits by cloning the Mount Baekdu tiger.[27] At the time, this professor was elevated to celebrity status and books described him as a potential Nobel Prize winner. Unfortunately, these hopes were premature. Korea, as of 2018, has yet to win a Nobel Prize in a scientific field from any national university or business affiliated researcher.

This professor first became known for his work in cloning animals, especially cattle and the dog, "Snuppy." Thereafter he published work on

---

[26] Savulescu, J. (2008, March 01). Research ethics and lessons from Hwanggate: What can we learn from the Korean cloning fraud? Retrieved from https://jme.bmj.com/content/34/3/214?legid=medethics;34/3/214

[27] Savulescu, J. (2008, March 01). Research ethics and lessons from Hwanggate: What can we learn from the Korean cloning fraud? Retrieved from https://jme.bmj.com/content/34/3/214?legid=medethics;34/3/214

stem cells, including landmark papers in 2004 and 2005 that brought him fame. The implications of these stem-cell publications meant that the group had cloned a human embryo and produced stem cells from it, a new possibility for individualized medicine and revolutionary in nature.[28] This was short-lived excitement, in 2004, questions were raised about donor egg procurement and in 2005, anonymous sources noted that these donor eggs were the result of unethical coercion from laboratory employees. An American professor who co-authored the 2005 paper attempted to remove his name because of doubts about research misconduct in this professor's lab.

In retrospect, the circumstances were ripe for misconduct by anyone thrust into such an important role. Linking his research to nationalism was a precarious way of promoting science, as any disappointing results would surely lead to negative public backlash. Extremely large amounts of money hinged on the successful outcomes of this professor's research. This professor described himself as a workaholic who had extremely ambitious goals for himself and the government had very high expectations of him.[29] Some have stated perhaps he simply attempted to achieve unrealistic results within an unrealistic time frame, as his research in stem cells was very demanding

---

[28] Stewart, C. N. (2011). *Research ethics for scientists: A companion for students*. Oxford: Wiley-Blackwell.
[29] Stewart, C. N. (2011). *Research ethics for scientists: A companion for students*. Oxford: Wiley-Blackwell.

but required speedy breakthroughs that simply were not feasible.[30] It is possible that this professor had pressure to deliver due to irresponsible reporting and false hopes or misconceptions.[31] As such, it was a steep drop for this professor from the summit of scientific achievement when it became evident that this professor had committed research misconduct and potentially outright fraud in two articles describing stem cell lines derived from cloned human embryos.[32] Both papers were found to have contained fabricated data, notwithstanding the separate issue arose regarding the way this professor had procured donor eggs for the experiments.[33] Eventually the two papers were retracted based on these egregious ethical mistakes, which were broadcast internationally and described as one of the most disappointing cases of misconduct in the history of science. In 2006, President Chung Un-Chan of Seoul National

---

[30] Stewart, C. N. (2011). *Research ethics for scientists: A companion for students*. Oxford: Wiley-Blackwell.

[31] Savulescu, J. (2008, March 01). Research ethics and lessons from Hwanggate: What can we learn from the Korean cloning fraud? Retrieved from https://jme.bmj.com/content/34/3/214?legid=medethics;34/3/214

[32] Kennedy, D. (2006, January 06). Editorial expression of concern. Retrieved from https://www.ncbi.nlm.nih.gov/pubmed/16373531; Kennedy, D. (2006, January 20). Editorial retraction. Retrieved from https://www.ncbi.nlm.nih.gov/pubmed/16410485

[33] Cyranoski, D. (2014). Cloning comeback. *Nature, 505*(7484), 468-471. doi:10.1038/505468a

University called the incident "an unwashable blemish on the whole scientific community as well as our country."[34]

A university investigation committee found that this professor's human-cloning papers were fraudulent, and both were later retracted by *Science* in January 2006. Essentially, images and graphs from the 2004 and 2005 publications were fabricated in order to make normal embryonic stem-cell lines and a cell line that underwent parthenogenesis appear as though they were cloned. The stem cells lacked DNA matches, meaning that no cloning had occurred. The investigation committee stated that, "[this professor] does not possess patient-specific stem-cell lines or any scientific bases" to think otherwise.[35] In March 2006, this professor was expelled from his university and the police authorities raided his laboratory in the launch of a massive investigation.[36] That same month, this professor admitted that he had directed researchers in his laboratory to fabricate data for the 2005 fraudulent paper.[37]

---

[34] Cyranoski, D. (2014). Cloning comeback. *Nature, 505*(7484), 468-471. doi:10.1038/505468a
[35] Cyranoski, D. (2014). Cloning comeback. *Nature, 505*(7484), 468-471. doi:10.1038/505468a
[36] Cyranoski, D. (2014). Cloning comeback. *Nature, 505*(7484), 468-471. doi:10.1038/505468a
[37] Savulescu, J. (2008, March 01). Research ethics and lessons from Hwanggate: What can we learn from the Korean cloning fraud? Retrieved from https://jme.bmj.com/content/34/3/214?legid=medethics;34/3/214

Finally, 2009, this professor was convicted of fraud in a criminal court case because of the use of government funding in the misconduct, and given a two-year suspended prison sentence.[38]

Some observers have speculated that this professor was perhaps overeager, and in too much of a rush, to adequately conduct his research and obtain a publishable result because he believed that even if he did not have real evidence at the time the 2005 paper was submitted, he would probably have such evidence soon after submitting it.[39] However, this professor's laboratory has never been able to produce evidence to verify the retracted publication about cloned stem-cell lines. Even more tragic, is the fact that there may not even have been a need for this professor to engage in data fabrication: according to the university investigation, this professor was able to produce cloned human blastocysts. This result, along with the parthenogenesis outcome for the derived embryo, would have been publishable results in and of themselves.[40]

---

[38] Stewart, C. N. (2011). *Research ethics for scientists: A companion for students*. Oxford: Wiley-Blackwell.
[39] Savulescu, J. (2008, March 01). Research ethics and lessons from Hwanggate: What can we learn from the Korean cloning fraud? Retrieved from https://jme.bmj.com/content/34/3/214?legid=medethics;34/3/214
[40] Savulescu, J. (2008, March 01). Research ethics and lessons from Hwanggate: What can we learn from the Korean cloning fraud? Retrieved from https://jme.bmj.com/content/34/3/214?legid=medethi

This professor and the media coverage left supporters in disbelief, and patients devastated. This professor had promised a patient, paralyzed as a result of a car accident, that she would someday walk again. The patient remained paralyzed until her death in 2010.[41] Curiously, this professor has attempted to rehabilitate his image within the scientific community after the misconduct. *Nature*, after publishing an update on the ten-year anniversary follow-up on this professor and the media coverage, even went the additional step of publishing a guidance article that clarified that this professor was still a *persona non grata* in the research ethics community.[42] Some people may believe that this professor has paid his dues and bid farewell to the practices that left him ostracized from the stem-cell cloning community. However, such ideas are dangerous because this professor has repeatedly refused to accept responsibility for his past actions. This professor has argued that his cells were the first cloned human stem-cell line, and ignored findings from the SNU investigation and retracted publications. This professor has even sought to get non-scientific legal judgments and court opinions to validate his wrongful assertions that contradict the fabricated

---

cs;34/3/214

[41] Cyranoski, D. (2014). Cloning comeback. *Nature, 505*(7484), 468-471. doi:10.1038/505468a

[42] Don't rush to rehabilitate Hwang. Retrieved from https://www.nature.com/news/don-t-rush-to-rehabilitate-hwang-1.14554

data from his own laboratory. The issue here is that this professor's fraud has actually set back Korean biotechnology research, rather than advance it. Korean researchers in this area of research and others have been viewed with enhanced scrutiny due to the global awareness of his misconduct. Instead of a global leader in this discovery, this professor has actually made it more difficult for researchers in this area. Meanwhile, those who brought this professor's misconduct to light were met with criticism and harassment, portrayed as traitors to Korea. This is the opposite of the reality, and such blowback against whistleblowers reflected the dangers when scientists attempt to use personal arrogance, overzealous haste, and national pride as a supplement for the ethical conduct of research.

When a scientist is found, conclusively, to have engaged in cheating such as data fabrication, it is common for the accused to claim it was merely a mistake, to apologize, and then explain that the benefits have outweighed the seemingly minor error. In other cases, researchers will claim to have simply forgotten or misplaced the data. Oftentimes, this is done with an eye towards the time saved from having to actually conduct the non-fabricated experiments.[43] To be clear, these are all poor excuses for data fabrication. Data fabrication has ramifications for the entire scientific community that cascade far beyond the mere individuals

---

[43] Stewart, C. N. (2011). *Research ethics for scientists: A companion for students*. Oxford: Wiley-Blackwell.

directly involved. As such, it is somewhat unusual that this professor continues to conduct research in Korea. Normally, researchers who engage in misconduct are barred from receiving governmental funding and affiliating with institutions of research. Journals often blacklist individuals who have been found guilty of data fabrication resulting in multiple retracted papers. Korea must do more to democratize and diversify its system of funding mechanisms in order to avoid promoting oligopolies where only a handful of academics receive the majority of funding.[44] Reserving grants and financial rewards exclusively for ethical researchers would also help clean up the current state of scientific research in Korea.[45] Professor Myung-Hee Chung of Gachon University was the Chair of Seoul National University's Investigation Committee in 2006.[46] According to him, the same types of research misconduct have

---

[44] Hong, S. (2016). We Should Not Forget Lessons Learned from the Woo Suk Hwang's Case of Research Misconduct and Bioethics Law Violation. *Journal of Korean Medical Science, 31*(11), 1671. doi:10.3346/jkms.2016.31.11.1671

[45] Hong, S. (2016). We Should Not Forget Lessons Learned from the Woo Suk Hwang's Case of Research Misconduct and Bioethics Law Violation. *Journal of Korean Medical Science, 31*(11), 1671. doi:10.3346/jkms.2016.31.11.1671

[46] Hong, S. (2016). We Should Not Forget Lessons Learned from the Woo Suk Hwang's Case of Research Misconduct and Bioethics Law Violation. *Journal of Korean Medical Science, 31*(11), 1671. doi:10.3346/jkms.2016.31.11.1671

taken place in the aftermath of the incident, and blames the biased and political governmental grant funding system, which concentrates grants in small groups who have abundant funds while other researchers have financial obstacles to research.[47]

Perhaps second only to this professor in notoriety, the case of a German physicist also merits analysis. An experimental physicist for a corporate research institute in the United States, the German physicist and co-authors published several well-received papers on changing non-conducting, organic molecules into semiconductors. Interestingly, the physicist's publication rate was a new finished paper every two weeks—a hyper-prolific record of accomplishment.[48] Until his fall in 2002, he had routinely won major awards and prizes for his scientific achievements. However, this suddenly changed after allegations surfaced that his experiments were not capable of being independently verified by reproduction or replication in other labs following his same methods outlined in the publications. A closer inspection of these papers showed new anomalies, such as images repeated across different publications. The physicist also claimed to be missing lab notebooks and blamed a computer

---

[47] Hong, S. (2016). We Should Not Forget Lessons Learned from the Woo Suk Hwang's Case of Research Misconduct and Bioethics Law Violation. *Journal of Korean Medical Science, 31*(11), 1671. doi:10.3346/jkms.2016.31.11.1671

[48] Stewart, C. N. (2011). *Research ethics for scientists: A companion for students*. Oxford: Wiley-Blackwell.

crash on missing digital copies. He was fired in 2002 after an investigation found him guilty of 16 cases of misconduct.[49]

It bears worth repeating that data fabrication is possible alongside image fabrication. Many times, repeated figures or charts, such as in the cases of the Korean professor and the German physicist, will reveal underlying fabricated data. When dealing with images, a researcher should be careful not to mislead a potential reader of a publication or grant proposal. Fabricated images could include copying and pasting images together to represent a single experiment, changing contrast levels and post-processing on images in software packages, concealing information to hide multiple interpretations, removing file information, and cropping images. All of these can also relate to experiments that were never actually performed. Non-linear corrections such as a gamma correction are almost never appropriate and should be disclosed when necessary.[50] If in doubt, a researcher should disclose. Prior to the media coverage on the Korean professor, some journals had added employees and safeguards to protect themselves from such possible fabrication: the *Journal of Cell Biology* has full-time staff to review

---

[49] Stewart, C. N. (2011). *Research ethics for scientists: A companion for students*. Oxford: Wiley-Blackwell.
[50] Rossner, M., & Yamada, K. M. (2004). Whats in a picture? The temptation of image manipulation. *The Journal of Cell Biology, 166*(1), 11-15. doi:10.1083/jcb.200406019

submitted images for digital manipulations.[51] Since then, software is becoming increasingly sophisticated to detect such image ethical issues.

[51] Gottweis, H., & Triendl, R. (2006). South Korean policy failure and the Hwang debacle. *Nature Biotechnology, 24*(2), 141-143. doi:10.1038/nbt0206-141

9. Falsification

Falsification, as stated previously, includes manipulating research materials, equipment, data, results so that the research is misrepresented in an inaccurate way. This can occur in a variety of ways.

For instance, a famous American university fired a promising immunology researcher and biologist after he fabricated and altered research papers in order to support grant applications.[52] The pressure to publish can result in incentives for morally wayward researchers to falsify their data in publications and grant submissions. Some researchers begin the process of submitting publications to journals even before they can formulate valid tests and assess preliminary findings in light of their goals, with the hope that they will successfully predict the outcome very quickly before the paper is published. The failure to wait until a study is completed can result in falsified data that attempts to prove an outcome or result from the experiment that has yet to be completed, and is known as a falsified, fraudulent act of misconduct. Taking into account our own cultural biases, we may be familiar with this risk.

---

[52] McCook, A. (2017, April 27). Harvard teaching hospital to pay $10 million to settle research misconduct allegations. Retrieved from https://retractionwatch.com/2017/04/27/harvard-teaching-hospital-pay-10-million-settle-research-misconduct-allegations/

Returning to the example of the Korean professor, it is worth mentioning that this professor was under extreme pressure due to his rock star status in Korea at the time of his misconduct. According to Hak Soo-Han, a producer at Munwha Broadcasting Corporation's PD Notebook, this was not necessarily the case: although "[p]eople say Koreans are expected to do everything fast," he notes, "[t]his was [this professor's] personal desire."[53]

Not only study results, but also patient data can be affected by acts of falsification. A research coordinator at an American hospital, changed patient medical records to make it easier for them to enroll in cancer drug trials. The patients, one of whom died, never should have been admitted to the study due to pre-existing conditions such as kidney liver function issues. The research coordinator was convicted of criminally negligent homicide and sentenced to 71 months in prison.[54]

Dealing with blotting can be a concern for clinical researchers and image falsification. In a different example from Korea, a biochemistry professor at had three articles retracted and over ten corrected for image manipulation, duplication, or errors.[55] In the *Journal of Biological Chemistry*,

---

[53] Cyranoski, D. (2006). Rise and fall. *News@nature*. doi:10.1038/news060109-8

[54] Fraud earns researcher time in jail. (n.d.). Retrieved from https://www.the-scientist.com/news-analysis/fraud-earns-researcher-time-in-jail-48058

the retracted paper featured duplications of an immunoblot across multiple lanes and multiple bands: “The HIF-1α immunoblot on the left in Fig. 1B was reused in Fig. 8A.”[56] Additionally, lanes were copied and pasted on top of other immunoblots, some of which contained more than 8 lanes of data. Similarly, a Brazilian researcher, in explaining duplicated and unusual western blot figures in a publication, noted factors such as a large number of students and researchers in the laboratory, large numbers of western blots produced by students, western blot registries based on radiograph scanning by different students, and electronic archives that could be changed by third parties on publicly hosted access points.[57] In a more intentional case, another former professor at

---

[55] Oransky, I. (2018, November 20). Researcher in South Korea racks up three retractions and at least 10 corrections. Retrieved from https://retractionwatch.com/2018/11/20/researcher-in-south-korea-racks-up-three-retractions-and-at-least-10-corrections

[56] Oransky, I. (2018, November 20). Researcher in South Korea racks up three retractions and at least 10 corrections. Retrieved from https://retractionwatch.com/2018/11/20/researcher-in-south-korea-racks-up-three-retractions-and-at-least-10-corrections

[57] Oransky, I. (2018, October 25). Three more retractions brings diabetes researcher who once sued publishers to 18. Retrieved from https://retractionwatch.com/2018/10/26/three-more-retractions-brings-diabetes-researcher-who-once-sued-publishers-to-18/#more-75381

a different American university falsified western blot data in publications by flipping them horizontally across the y-axis to represent both 2 high glucose and 1 peroxynitrite treatments.[58]

A similar instance of blotting falsifications and misconduct involves a Taiwanese anticancer expert and professor who was found to have committed research fraud beginning in 2001 and throughout his almost 20-year career at an American university.[59] The expert likely cropped and pasted lanes onto a blot, the measure of protein expression and changes, in order to make it show a constant and equal protein expression.[60] The expert manipulated images, reused blots, and pasted lines on images published to support data and serve as evidence for positive findings. In another case, the expert admitted that he changing images by falsifying the particular conditions and

---

[58] NIH. NOT-OD-18-006: FINDINGS OF RESEARCH MISCONDUCT. Retrieved from https://grants.nih.gov/grants/guide/notice-files/NOT-OD-18-006.html

[59] Salmonsen, R. (2018, April 01). Former Taiwan Academia Sinica professor resig... Retrieved from https://www.taiwannews.com.tw/en/news/3395539

[60] Cartwright. Ohio State researcher forced to resign after falsifying data in cancer research projects paid for in part by Pelotonia and Stefanie Spielman cancer fund. Retrieved from https://www.thelantern.com/2018/03/ohio-state-researcher-forced-to-resign-after-falsifying-data-in-cancer-research-projects-paid-for-in-part-by-pelotonia-and-stefanie-spielman-cancer-fund/

situation of the experiment, after the expert became frustrated with a student researcher's lack of speed in completing the work.[61] The expert falsified data on compounds that were intended for use on patients, thus directly endangering the health and safety of patients. Despite blaming data issues on postdoctoral researchers and other lab workers, the expert was found guilty of falsification because of computer records: his hard drive contained timestamps showing when images were altered, including raw files that no one else possessed. Since no one else in the lab where the expert worked had saved records, and no notebooks were kept in violation of state and federal government policy, this was conclusive in proving misconduct.[62] Similarly, Adobe Photoshop and other image editing programs have been used to attempt to manipulate images.[63]

---

[61] Cartwright. Ohio State researcher forced to resign after falsifying data in cancer research projects paid for in part by Pelotonia and Stefanie Spielman cancer fund. Retrieved from https://www.thelantern.com/2018/03/ohio-state-researcher-forced-to-resign-after-falsifying-data-in-cancer-research-projects-paid-for-in-part-by-pelotonia-and-stefanie-spielman-cancer-fund//

[62] Cartwright. Ohio State researcher forced to resign after falsifying data in cancer research projects paid for in part by Pelotonia and Stefanie Spielman cancer fund. Retrieved from https://www.thelantern.com/2018/03/ohio-state-researcher-forced-to-resign-after-falsifying-data-in-cancer-research-projects-paid-for-in-part-by-pelotonia-and-stefanie-spielman-cancer-fund/

In a related example, a researcher falsified data from the National Institute of Alcohol Abuse and Alcoholism and the National Institutes of Health. The researcher changed bar graphs by up to 50% and falsified other data to exaggerate alcohol exposure's effect on fetal development and to make experimental results seem more significant, positive, and consistent with his hypothesis that ethanol exposure in-utero alters normal brain development in a theorized manner.[64] After laboratory workers came forward with allegations of misconduct, an investigation revealed computer files and lab notes that corroborated accusations. In essence, the researcher had committed substantial fraud in multiple instances to make the data lead to where he wanted it to lead. Some questioned why figures contained the same actin blot, and it should be noted that some of these had the explanation that they had been reprobed with a mouse monoclonal anti-actin antibody after being stripped of the immunolabel.[65]

---

[63] UConn Investigator. Retrieved from https://www.chronicle.com/blogs/percolator/uconn-investigation-finds-that-health-researcher-fabricated-data/28291

[64] Oransky, I. (2012, February 26). ORI: Former SUNY Upstate neuroscience dept. chair Miller manipulated data in four grant applications. Retrieved from https://retractionwatch.com/2012/02/26/ori-former-suny-upstate-neuroscience-dept-chair-miller-manipulated-data-in-four-grant-applications/

[65] Oransky, I. (2012, February 26). ORI: Former SUNY

Upstate neuroscience dept. chair Miller manipulated data in four grant applications. Retrieved from https://retractionwatch.com/2012/02/26/ori-former-suny-upstate-neuroscience-dept-chair-miller-manipulated-data-in-four-grant-applications/

10. Plagiarism

Plagiarism is the appropriation, or taking, of the words, ideas, processes, results, or related concepts of another without proper attribution and giving appropriate credit. It means claiming credit for the ideas, sentences, or phrases of another.[66]

In a research context, this could mean using the scientific advances or potential discoveries of another to claim credit for oneself. In this way, it appears that the author did all the work that resulted in the data. In other cases, writer's block, or the inability to easily formulate and record thoughts in written form, leads to poor choices by researchers in stealing content from others.

Scientific publications have different sections. Writing each of these sections requires different skills. The abstract, for instance, may be the first part of a paper that we read. However, often the abstract is actually the last part of a manuscript to be written, as it summarizes and explains the reasons and takeaways from the paper. Afterwards is the introduction, which, is distinct from the abstract in that it may provide an overview of commonly held ideas and background information on theories or ideas the manuscript attempts to build upon. This is a potential section that plagiarism may be more likely to occur. The

---

[66] Stewart, C. N. (2011). *Research ethics for scientists: A companion for students*. Oxford: Wiley-Blackwell.

introduction is followed by the materials and methods section, which may be the most at risk section of a scientific manuscript for plagiarism. Here the author describes the specific technical steps and processes that allow the experiment to occur. This may involve repeated phrases and technical terms for which no synonyms exist. In particular, the names of chemicals, specific formulas, geographic descriptions and other similar methods make up common language for subsets of the research area.[67] Revising the materials and methods section can be difficult, especially for those who do not hold fluency in the target language.[68] It can be tempting to omit things from the methods section, but this tendency conflicts with the goal of scientific writing to convey and communicate about learned knowledge and past actions.[69] The methods section should be carefully evaluated when writing to avoid plagiarism for these reasons. Finally, the results section and the discussion section complete the paper format. An appendix or supplement will likely also include tables, figures, citations, or words of gratitude and acknowledgement to helpful contributors.[70] For the medical sciences, publications often rely on paraphrasing and citations rather than direct

---

[67] Stewart, C. N. (2011). *Research ethics for scientists: A companion for students*. Oxford: Wiley-Blackwell.
[68] Stewart, C. N. (2011). *Research ethics for scientists: A companion for students*. Oxford: Wiley-Blackwell.
[69] Stewart, C. N. (2011). *Research ethics for scientists: A companion for students*. Oxford: Wiley-Blackwell.
[70] Stewart, C. N. (2011). *Research ethics for scientists: A companion for students*. Oxford: Wiley-Blackwell.

quotations. It would be rare to see a lengthy direct quotation in a leading medical journal. This differs from other scientific writing practices in other fields, and can make plagiarism more or less noticeable depending on the degree and quality of paraphrasing that has occurred.

When paraphrasing a sentence, it is important to change the underlying structure of sentence meant to be adopted into the new work, not merely changing grammatical tenses, swapping words for synonyms, or rearranging the order of subordinate clauses and commas.

Plagiarism detection software has increasingly come to be used to scan potential manuscripts before publication. Universities, such as Seoul National University, also possess multiple different packages to review student and faculty writing. These programs measure not only exact matches against a database of papers, but also feature similarity indexes and other means of evaluating resemblance to other works.

Plagiarism can commonly be detected by looking at citations in papers. If a source is referenced in the manuscript, but not listed among the complete list of sources, it may be an incidence where the author has copied the attribution from another paper but failed to copy the list of works cited. Other times, plagiarized sentences may have differences in quality or tone that are easily detected by those familiar with the literature and language of scientific publishing. For a number of different reasons, plagiarism has been highly visible in recent years and traces back earlier.

For instance, in 1991 an American professor was found to have plagiarized from an article he had reviewed as a peer reviewer. Not simply committing conflict of interest in violation of the duty of confidentiality as the peer reviewer had included in the introduction section of a government grant proposal information from an unpublished article for which he was serving as peer reviewer in the *Journal of the American Chemical Society*. The professor did not cite or attribute this information to the unpublished article's author.[71] The peer reviewer, included text from the confidential grant application for which Paquette was a reviewer, in his own subsequent grant application.[72]

Plagiarism can also combine with fabrication or falsification to form especially toxic forms of misconduct in research. Plagiarism-fabrication, also known as plagiofabrication or poaching, involves taking an irrelevant, unrelated picture, figure, or graphic from a random, unrelated publication and submitting it as relevant to the experiment and a representation of new

---

[71] Ohio State University. Ohio State News. Retrieved from https://news.osu.edu/scientific-misconduct-charge-ruled-valid/

[72] NIH. NOT-93-177: FINAL FINDINGS OF SCIENTIFIC MISCONDUCT. Retrieved from https://grants.nih.gov/grants/guide/notice-files/not93-177.html

data. It can be very difficult to detect this type of plagiarism, because images from old publications may look perfect to everyone except for the original author of the plagiarism image.[73] A leukemia researcher at a Spanish university has appropriated images from the work of others for his published papers as a pattern of routine behavior, taking entire images from the papers of at least three different groups.[74] A researcher at a German university found that two former postdoctoral researchers committed plagiarism-fabrication when they used protein blots from unrelated experiments to justify findings on cell signaling for their work on allergic reactions and asthma.[75]

Self-plagiarism can be a difficult concept to comprehend for researchers newly introduced to ethics. Multiple publication of the same results or

---

[73] Marcus, A. (2013, February 08). Lifted figure prompts retraction of Oncogene paper by Roman-Gomez. Retrieved from https://retractionwatch.com/2013/02/08/lifted-figure-prompts-retraction-of-oncogene-paper-by-roman-gomez/

[74] Marcus, A. (2013, February 08). Lifted figure prompts retraction of Oncogene paper by Roman-Gomez. Retrieved from https://retractionwatch.com/2013/02/08/lifted-figure-prompts-retraction-of-oncogene-paper-by-roman-gomez/

[75] Schiermeier, Q. (2010). German research centre widens misconduct probe. *Nature*. doi:10.1038/news.2010.671

the same content with only slightly different title, or with the same title but in different journals, is often considered misconduct. For some editors, publishing the same manuscript in a different language may also constitute misconduct. Many journals have publication submission policies that ask whether this manuscript was previously published. Because one of the basic assumptions about a submitted paper is that it has not appeared as a publication somewhere else before, these types of self-plagiarism constitute abuse of the scientific publishing process.

Sometimes it may be acceptable or permissible to segment or divide a study into two separate works. For instance, if a researcher wishes to publish a standard academic journal article about a study, in addition to a more popular media type of piece such as a magazine study, this would probably not be an ethical violation as long as the researcher discloses both publications to both publishers.[76]

Additionally, some sources differentiate between text recycling and self-plagiarism. In text recycling, text that has been used before is reused by the author for a new work. For instance, an author may want to use writing from an

---

[76] Dealing with Self-Plagiarism (a.k.a. Text Recycling) in Law Reviews. (2017, July 26). Retrieved from https://ripslawlibrarian.wordpress.com/2017/08/01/dealing-with-self-plagiarism-a-k-a-text-recycling-in-law-reviews/

unpublished manuscript that they have no intention of ever publishing and add it to another document that may or may not be published. In this situation, recycling the text would likely be acceptable. However, the inquiry focuses on whether or not the author has reused their own work and whether the reusable work has been published before. If the author is doubtful, it would be better to disclose that this text had appeared earlier. For instance, if a researcher wants to write a book chapter for a textbook, they may rely on a published manuscript that formerly appeared in an academic journal. Here, the researcher should disclose to the textbook that they are basing material on the former article. The textbook may wish to add a disclaimer, general disclosure, or footnotes to attribute the material and avoid text recycling issues.

Cryptoamnesia is another concern for young researchers new to the ethical side of science. In this subset of plagiarism, a writer usually inadvertently or unintentionally uses the words of another scientist that has been learned earlier, sometimes weeks, months, or years beforehand.[77]

Aside from ensuring that authors receive adequate compensation and credit for their work,

---

[77] Pecorari, D. (2003). Good and original: Plagiarism and patchwriting in academic second-language writing. *Journal of Second Language Writing, 12*(4), 317-345. doi:10.1016/j.jslw.2003.08.004

plagiarism can also have implications for intellectual property rights. When an academic article is published in a journal, typically the author will transfer publication rights to the journal. This allows the journal to actually own the ability to print the article and use the author's words. In legal terms, the article is copyrighted. This concept, copyright, helps explain why publishing the same images in two different papers is not allowed. Not only are duplicated images a waste of time and effort for those seeking new, innovative research, it also violates this copyright to reprint pictures without permission from the copyright holder such as the prior journal or authors who published the image originally.[78]

Patents are another form of intellectual property right, but distinct from copyright in some ways. Patents are based on the inventorship of a new "process, machine, manufacture, or composition of matter," such as a medical device or a formulation of a new drug, while copyright based on the authorship of a new expression of idea (in American patent law, 35 U.S. Code § 101). When writing a patent for the United States Patent and Trademark Office, it is acceptable for a patent application to contain material that has been published by the author previously, even if a patent attorney is submitting the patent application for the inventor.[79] However, the potential patent

---

[78] Stewart, C. N. (2011). *Research ethics for scientists: A companion for students*. Oxford: Wiley-Blackwell.
[79] Check, E. (2006). Journals scolded for slack disclosure

application must also be disclosed to a journal that requests it. For instance, *Science* requires all authors who submit papers to the journal to disclose whether they plan to submit or have already been granted a patent based on this work described in the article.[80] *Nature* has a similar submission question that asks authors the same, as well as if patents they own may be harmed or affected by the publication.[81] These questions are important to assess whether publication could make it easier for other persons to challenge the inventorship of the patent. Such challenges may be based on potential pieces of evidence, such as laboratory notebooks or records.. In summary, intellectual property also justifies a strict assessment of plagiarism in its many forms as research misconduct.

> Seoul National University Research Ethics Guidelines
>
> Revised on July 16, 2010.
>
> Chapter 2 Integrity in Research Presentation
>
> Article 8
>
> 【Use of One's Own Research Outcome】
>
> ① In preparing research literature, a researcher shall use, in principle, his/her own

---

rules. *News@nature*. doi:10.1038/news060116-6

[80] Check, E. (2006). Journals scolded for slack disclosure rules. *News@nature*. doi:10.1038/news060116-6

[81] Check, E. (2006). Journals scolded for slack disclosure rules. *News@nature*. doi:10.1038/news060116-6

research idea, research data, and sentences and include research results not yet presented before.

② In preparing the research literature or a research proposal, a researcher may partially use his/her own research results that have already been published within the scope that does not affect the originality of such research. However, such research data shall be used with correct indication of its source and shall not be suggested as if it were presented for the first time in such research. In cases wherein no less than one paragraph or five sentences are consecutively reused from these prepared in the past, correct indication of its source or citation shall be made.

③ In cases wherein his/her own research outcomes that have already been presented are inserted into a textbook and currently used as general knowledge, the researcher may use all or part of such research outcome without indication of its source and citation.

Seoul National University Research Ethics Guidelines

Revised on July 16, 2010.

Chapter 2 Integrity in Research Presentation

Article 7

【Use of Research Outcome by Other Persons】

① In preparing the research literature or a research proposal, a researcher shall, in principle, express the contents thereof in his/her own sentences based on in his/her own research idea or research data.

② In preparing the research literature or a research proposal, a researcher shall not use the research outcome by any other person and which has already been presented or published (including unpublished outcomes such as research proposal or manuscript) in its original form or other transformed form as if it were his/her own research outcome.

③ In preparing the research literature or a research proposal, a researcher may partially use the research idea, research data, and sentences by any other person within the scope that does not affect the originality of his/her own research. However, in such case, correct indication of its source or citation shall be made, with the researcher refraining from engaging in any of the following acts:

1. Act of expressing all or any part of the research idea and research data by any other person using a different way of description as if it were his/her own research outcome;

2. Act of using the sentences written by any other person as if they were his/her own sentences (case of using two or more sentences written by other persons without indication of source or citation shall be presumed to correspond to such act, and its final judgment shall be made considering the characteristics of a major field of study

11. Investigations and Whistleblowers

As seen in previous sections, there is an established hierarchy and way of investigating accusations of unethical research. This involves verifying the people involved in each situation, writing reports, and making determinations about what occurred and how it might be prevented in the future.

Many universities, such as Seoul National University, have offices dedicated to research integrity. Anonymous reporting opportunities and hotlines can provide a way for whistleblowers to detail potential situations worthy of investigation. Sometimes, the best course of action is to confidentially discuss any potential issues with a trusted friend or neutral advisor. At the same time, should a tip or report result in an administrative hearing, the person who initially reported the conduct should be prepared to testify. Not only can unethical behavior be a form of research misconduct, but the failure to report unethical behavior can also be a form of research misconduct. Given Korea's cultural emphasis on harmonious relations, misconduct by a single member of a group can bring shame to all the members collectively.[82] Honest individual researchers who practice ethically upright research may be overshadowed by infamous bad actors who poison the reputation of their group and harm the image of scientific research. When misconduct is revealed, it often comes from those who are close to the discovery in the roles of journal reviewer, journal editor, lab partner, or even journal readers. This relationship dynamic can make it extremely difficult for whistleblowers in Korea to publicize cases of missing or fabricated data. After all, it would be difficult to accept a letter of recommendation from a former, discredited laboratory supervisor.

In this regard, the need to point out and make known suspected unethical research can be a delicate process. Often, when concerns about

---

[82] Nho, H. (2016). Research ethics education in Korea for overcoming culture and value system differences. *Journal of Open Innovation: Technology, Market, and Complexity, 2*(1). doi:10.1186/s40852-016-0030-3

potential misconduct or other behavior reaches a critical mass, it is up to those who may be close to or working alongside the suspected plagiarist, fabricator, or falsifier to report. At the same time, these potential whistleblowers have their own decisions to make in whether to stay silent or to speak up. However, reporting unethical research can have consequences both for the whistleblower and for the suspected researcher. Increasingly, anonymous watchdogs are turning to the internet, email, and media to publicize these concerns.[83]

In Korea, this can be an especially potent form of raising awareness about pervasive failures in ethical management, whether for research or otherwise. It is worth it pointing out that Korea has specific values regarding the centrality of relationships that potentially make internal whistleblowing difficult to function properly. Specifically, rigid hierarchies may make it difficult to form relationships based on equality between professors and students, thus prohibiting open and honest communication.[84] Given this dilemma, resorting to neutral, internal arbiters together with outside resources can help mitigate this problem.

---

[83] Yong, E., Ledford, H., & Noorden, R. V. (2013). Research ethics: 3 ways to blow the whistle. *Nature, 503*(7477), 454-457. doi:10.1038/503454a

[84] Nho, H. (2016). Research ethics education in Korea for overcoming culture and value system differences. *Journal of Open Innovation: Technology, Market, and Complexity, 2*(1). doi:10.1186/s40852-016-0030-3

For instance, returning to the first example of the professor at a famous Korean university, both internal university stakeholders and outside resources combined to definitively discover and prove unethical conduct. Initially, much of the details about this professor were discussed on internet forums.[85] The exposure of this professor as a fraudulent researcher supports the manner in which the case was pursued: at the very beginning, those acting as good faith whistleblowers were protected through the anonymous medium of the internet. Although the expulsion of this professor required many online posts in order to gain credibility and build momentum, this was a benefit for whistleblowers in that it allowed them to prevent skeptical, bad-faith bloggers from poisoning or abusing the forum.[86] The internet allowed young scientists to scrutinize this professor's data sets of stained cell images and DNA fingerprints to post collaborative findings anonymously online.[87] The webpage for the Biological Research Information Center (BRIC)

---

[85] Savulescu, J. (2008, March 01). Research ethics and lessons from Hwanggate: What can we learn from the Korean cloning fraud? Retrieved from https://jme.bmj.com/content/34/3/214?legid=medethics;34/3/214

[86] Savulescu, J. (2008, March 01). Research ethics and lessons from Hwanggate: What can we learn from the Korean cloning fraud? Retrieved from https://jme.bmj.com/content/34/3/214?legid=medethics;34/3/214

[87] Cyranoski, D. (2006). Rise and fall. *News@nature*. doi:10.1038/news060109-8

featured as a base for these online sleuths to point out potential errors in the papers, which contributed to the opening of the internal investigation at the university. Rather than a negative, this idea to post data sets online has actually come to be seen as beneficial for most researchers. Utilizing the internet and public involvement to function as an additional layer of ethical oversight is commonly done when research protocols, patient information, and consent forms are removed of personal health identification information and posted online.[88] It is clear that anonymous whistleblowing will likely continue to become more common as more and more people globally have access to research and as tools to detect plagiarism and image manipulation become ever more sophisticated.[89] For instance, social media sites like Twitter have been used to report potential plagiarism to high profile journals such as *Nature Methods*.[90]

At the same time, resorting to anonymous reporting can evoke feelings of suspicion or mistrust in relationship-oriented cultures like those common in East Asia. Some editors worry that

---

[88] Cyranoski, D. (2006). Rise and fall. *News@nature*. doi:10.1038/news060109-8

[89] Yong, E., Ledford, H., & Noorden, R. V. (2013). Research ethics: 3 ways to blow the whistle. *Nature, 503*(7477), 454-457.

[90] NatureMaterials. (2013, August 08). Twitter. Retrieved from https://twitter.com/NatureMaterials/statuses/365690950162452481

unfounded allegations of scientific misconduct can do harm or be scientific conduct themselves if they attempt to false flag or discredit innocent actors.[91] Anonymous tip-offs can help compensate the inherent loss of trust from anonymity be being specific, detailed, and respectful.[92] Since 2010, a prolific anonymous whistleblower going by the pseudonym "Clare Francis" has sent hundreds of emails to editors of life-science journals to point out potential plagiarism, data fabrication, and other misconduct. It is unknown who the pseudonym actually is, whether an individual or a group of people. In any case, "Clare Francis" emails have resulted in numerous corrections and retractions. Although some editors viewed the notices as intrusive and excessive at times, it is clear that the anonymous whistleblower is an overall force for ethical research. In the wake of "the Clare Francis phenomenon" the debate about anonymous whistleblowing continues, as more websites allow people to publicly detail issues or complaints about research papers.[93]

---

[91] Oransky, I. (2013, December 06). Should scientific misconduct be handled by the police? It's fraud week at Nature and Nature Medicine. Retrieved from https://retractionwatch.com/2013/12/06/should-scientific-misconduct-be-handled-by-the-police-its-fraud-week-at-nature-and-nature-medicine/

[92] Yong, E., Ledford, H., & Noorden, R. V. (2013). Research ethics: 3 ways to blow the whistle. *Nature, 503*(7477), 454-457. doi:10.1038/503454a

[93] Yong, E., Ledford, H., & Noorden, R. V. (2013). Research ethics: 3 ways to blow the whistle. *Nature, 503*(7477), 454-457. doi:10.1038/503454

Returning to the Korean professor incident, a whistleblower joined this professor's laboratory at the famous Korean university in 2002, and revealed in 2013 that he was responsible for helping start the investigation that uncovered the fraud.[94] Since then, he has suffered from abuse and also received support from a sharply divided Korean public still reeling from the legacy of this professor, the former national hero. The whistleblower initially began to question his desire for publicity and left this professor's laboratory in 2004.[95] After this professor's 2005 paper[96] was published to worldwide acclaim in *Science*, the whistleblower became increasingly suspicious, noting, "I knew how difficult [the experiment] was," he says. "It wasn't logical."[97] The whistleblower attributes his desire to go public to his personal view of this alleged professor's personal shortcomings and claims to make the paralyzed walk again.[98] At the same time, the

---

[94] Cyranoski, D. (2014). Whistle-blower breaks his silence. *Nature, 505*(7485), 593-594. doi:10.1038/505593a

[95] Cyranoski, D. (2014). Whistle-blower breaks his silence. *Nature,505*(7485), 593-594. doi:10.1038/505593a

[96] Timeline of a controversy. (2005). *News@nature*. doi:10.1038/news051219-3

[97] Cyranoski, D. (2014). Whistle-blower breaks his silence. *Nature,505*(7485), 593-594. doi:10.1038/505593a

[98] Cyranoski, D. (2014). Whistle-blower breaks his silence. *Nature,505*(7485), 593-594.

whistleblower avoided internal whistleblower resources at the university, such as university research ethics outlets to report violations. Acting as an anonymous source, the whistleblower emailed Munwha Broadcasting Corporation (MBC) to elicit an investigation in June 2005. Thereafter a series of tumultuous investigations, first a program by MBC aired in November 2005, and later a forced confession from this professor. The ensuing public outcry postponed a second MBC program after 11 of 12 advertising sponsors withdrew from the TV network and producers faced threats of physical harm and lawsuits.[99] When the second program finally aired in December 2005, this professor's reputation had been damaged and the whistleblower's identity became known. Supporters of this professor protested at the gates of the university the findings, despite the data supporting the fact that the data was faked.[100] Sadly, this disclosure of the whistleblower's identity would likely not have occurred if the whistleblower had decided to use internal Seoul National University whistleblower resources. The whistleblower, his wife, his job, and family faced threatening emails from this professor's supporters and the whistleblower went into hiding for six months. The whistleblower resigned from his job

---

doi:10.1038/505593a

[99] Cyranoski, D., & Check, E. (2005). Clone star admits lies over eggs. *Nature, 438*(7068), 536-536. doi:10.1038/438536a

[100] Cyranoski, D. (2006). Rise and fall. *News@nature*. doi:10.1038/news060109-8

and remained unemployed until 2007.[101]

Unfortunately, ignorance on the part of the public and a lack of recognition of the role of ethical research in society resulted in criticism from this professor's supporters, who believed the whistleblower had harmed the Korean stem-cell industry and injured the nation's image with minor affairs.[102] These claims were categorically false, and the empirical evidence supports the notion that this professor's research when dealing with stem-cell research was flawed science, and this professor's work on cow cloning in 1990s yielded no scientific publications.[103] It is possible when looking back that the whistleblower did not cause Korea to lose anything, rather, the whistleblower benefitted Korea by helping to stop wasted efforts to promote dishonest and ineffective research.[104] This professor's mistaken efforts were stopped, and in doing so, Korean research could move on from the unfruitful branch of this professor's research.[105] Scholars have found that researchers

---

[101] Cyranoski, D. (2014). Whistle-blower breaks his silence. *Nature, 505*(7485), 593-594. doi:10.1038/505593a

[102] Cyranoski, D. (2014). Whistle-blower breaks his silence. *Nature, 505*(7485), 593-594. doi:10.1038/505593a

[103] Don't rush to rehabilitate Hwang. Retrieved from https://www.nature.com/news/don-t-rush-to-rehabilitate-hwang-1.14554

[104] Don't rush to rehabilitate Hwang. Retrieved from https://www.nature.com/news/don-t-rush-to-rehabilitate-hwang-1.14554

with multiple retractions are more likely to have additional publications retracted than those who have never had a paper retracted. Researchers with at least five publications retracted are very likely to have subsequent papers retracted.[106] Indeed, it is possible that if the whistleblower had stayed silent, this professor's fraudulent research could have continued for years before eventually being discovered and having an even greater harmful impact at that point in the future. As such, it is clear that whistleblowing carries some risks, especially for younger researchers but is a crucial step for systemic ethical reform.[107]

The whistleblower's example shows that whistleblowers can pay a price for their ethical sacrifice. Although as a society we believe that fraud should be punished and sacrifice rewarded, we are not entirely comfortable with the idea of discarding a relationship in order to report misconduct. To avoid this, some companies have hired individuals to act as whistleblowers and scout misconduct. Seoul National University also

---

[105] Don't rush to rehabilitate Hwang. Retrieved from https://www.nature.com/news/don-t-rush-to-rehabilitate-hwang-1.14554

[106] McCook, A. (2018, April 07). How often do scientists who commit misconduct do it again? Retrieved from https://retractionwatch.com/2018/04/09/how-often-do-scientists-who-commit-misconduct-do-it-again/#more-63896

[107] Cyranoski, D. (2014). Whistle-blower breaks his silence. *Nature, 505*(7485), 593-594. doi:10.1038/505593a

maintains a research integrity office to monitor these types of allegations. The internet, also, can act as a public conscience to guide morals, while at the same time being susceptible to groupthink and herd mentality. Some of these drawbacks to ethical research are specific to Korean society. In particular, safeguards for whistleblowers to avoid retaliation when reporting misconduct is important because of the concept of face-saving and maintaining a superficial image in Korean society. As the organization as a whole can suffer from misconduct, there may be a motivation to suppress findings of unethical research. Overall, a tendency to camouflage internal misconduct in order to preserve outer appearance can forestall any potential investigation of misconduct.[108] In especially deleterious situations, endemic unethical norms can result in those who adhere to ethical behavior and follow research ethics guidelines as being labeled foolish or naive by eschewing the potential for a quick result.[109]

In this sense, there is a type of gap between those ethical norms espoused by Western research

---

[108] Nho, H. (2016). Research ethics education in Korea for overcoming culture and value system differences. *Journal of Open Innovation: Technology, Market, and Complexity, 2*(1). doi:10.1186/s40852-016-0030-3

[109] Nho, H. (2016). Research ethics education in Korea for overcoming culture and value system differences. *Journal of Open Innovation: Technology, Market, and Complexity, 2*(1). doi:10.1186/s40852-016-0030-3

organizations and those actionable by Korean organizations. Researchers new to research ethics may question the extent of these types of whistleblowing and the usefulness of doing so. Traditional, Korean ethical standards may ask whether internal whistleblowing is permissible if a friend or mentor is suspected of misconduct: should someone allege misconduct by a friend or mentor? To this uncertainty, perhaps the example of an American professor and a co-author on the Korean professor's retracted publication, may prove useful. The American professor had previously described the Korean professor as his best friend and even as a brother.[110] Despite this, after learning of the Korean professor's suspected misconduct, the American professor decided to ask that his name be withdrawn from the publication they authored. However, again, the co-author has a responsibility to carefully review a draft manuscript to ensure that the entire work is within ethical standards. Withdrawing a name from the publication after publication is too little, too late. From the example of the American professor, another idea about whistleblowers is highlighted. The American professor working at the collaborating institution also prompted the investigation by his own university into the Korean

---

[110] Kolata, G. (2005, December 20). A Cloning Scandal Rocks a Pillar of Science Publishing. *New York Times*. Retrieved from https://archive.nytimes.com/www.nytimes.com/learning/teachers/featured_articles/20051220tuesday.html; Check, E. (2006). Schatten in the spotlight. *News@nature*. doi:10.1038/news060109-7

professor's misconduct. Because research is often international in nature, although internal whistleblowers can be disadvantaged by their own institutions, collaborative projects can have multi-lateral oversight that view Western ethical standards as a minimum prerequisite.[111]

Finally, the example of a former researcher at an American university, highlights the ways that cohort misconduct can reflect on otherwise innocent researchers. The former researcher was surprised by an impromptu misconduct investigation into his laboratory after some data in a publication were falsified. Though the former researcher had no knowledge of misconduct, several people in the laboratory were under investigation and no evidence emerged that he knew images or data were falsified.[112] The former researcher was victimized under selective prosecution and barred from receiving federal funds. Where guilty by association in a laboratory becomes a possibility, the need for whistleblower protections increases. When misconduct is

---

[111] Nho, H. (2016). Research ethics education in Korea for overcoming culture and value system differences. *Journal of Open Innovation: Technology, Market, and Complexity, 2*(1). doi:10.1186/s40852-016-0030-3

[112] Goldstein, R. (2018, August 13). Are you liable for misconduct by scientific collaborators? What a recent court decision could mean for scientists. Retrieved from https://retractionwatch.com/2018/08/13/are-you-liable-for-misconduct-by-scientific-collaborators-what-a-recent-court-decision-means-for-scientists/

discovered, journals act quickly to correct any errors. Authors accused of misconduct usually claim that there has been a mistake or confusion between files.[113] Regardless of intent, these types of manipulations result in retraction notices by journals. In extreme cases, accused researchers blame investigative committees for bias and expertise even when there is no basis to believe so.[114]

Scientists accused of wrongdoing have resorted to lawsuits for libel and defamation against media companies that publish information on suspected misconduct, often losing.[115] In a noteworthy case, a Canadian researcher who sued the Canadian Broadcasting Corporation for publicizing his research misconduct was ordered to pay $1.6 million in legal fees after the accusations of unethical conduct were found to be true.[116] Even the New York Times newspaper has faced such a now-dismissed lawsuit by a cancer researcher, whose

---

[113] Retractions Due to Image Manipulation by ESL Authors. (2018, May 23). Retrieved from https://www.enago.com/academy/retractions-due-to-image-manipulation-by-esl-authors/

[114] Cyranoski, D. (2006). Whos who. *News@nature*. doi:10.1038/news060109-9

[115] Palus, S. (2016, August 25). Nutrition researcher Chandra loses libel case against CBC. Retrieved from https://retractionwatch.com/2015/07/31/nutrition-researcher-chandra-loses-libel-case-against-cbc/

[116] Palus, S. (2016, August 25). Nutrition researcher Chandra loses libel case against CBC. Retrieved from https://retractionwatch.com/2015/07/31/nutrition-researcher-chandra-loses-libel-case-against-cbc/

allegations of misconduct warranted a front-page story.[117] These failed suits are a last resort of researchers unwilling to face the humiliation and pain of a tarnished career from unethical research practices.

Seoul National University Research Ethics Guidelines

Revised on July 16, 2010.

Chapter 1 Responsibility and Obligation of Researchers

Article 4

【Responsibility to Be Kept When Expressing Research Career】

① A researcher's own teaching career or research career shall be neither exaggerated, distorted, nor written in a fraudulent manner.

② Any research career that is unfavorable to a researcher himself/herself shall not intentionally be reduced or omitted for personal gain such as support for research expenses.

[117] Oransky, I. (2018, November 08). Judge dismisses most of Carlo Croce's libel case against the New York Times. Retrieved from https://retractionwatch.com/2018/11/08/judge-dismisses-most-of-carlo-croces-libel-case-against-the-new-york-times/#more-76922

12. Authorship

Authorship is an increasingly important part of scientific publication, and perhaps one of the most important benefits of research for scientific researchers.[118] Authorship determines the way that credit for scientific advances is assigned. As such, it is the main way that the scientific community evaluates the benefits received from contributions and innovations in developing the body of knowledge comprised of research. This has the simultaneous potential for benefit, but also a responsibility on the part of authors. Publications require actual contributions in order to claim authorship, and should demonstrate knowledge,

---

[118] Stewart, C. N. (2011). *Research ethics for scientists: A companion for students*. Oxford: Wiley-Blackwell.

effort, and expertise.[119]

Authorship requires a significant contribution to the project, whether that means conceptualization, design, data analysis, or a combination of these things. Additionally, when someone includes their name as an author on a publication, they signal that they will assume responsibility and ownership in the publication for their contributions. Notably, this does not only apply to the corresponding author or the first author, but rather to all authors listed. According to the United States National Institutes of Health, significant contribution to the research study is not enough, but also accompanies drafting or reviewing and revising the article itself. The International Committee of Medical Journal Editors makes the same recommendation, and goes a step further in requiring that final approval of the version to be published as a prerequisite for inclusion as an author. Additionally, merely sponsoring the laboratory where research is conducted, obtaining funding to start research, gathering data, and general supervision are usually not significant enough to merit contribution as an author. However, even if someone should not be included as a co-author, he or she can still be thanked in the paper. Usually those who have assisted with their advice and encouragement by providing an office, financial support, patient material, reagents, or other such items should be

---

[119] Stewart, C. N. (2011). *Research ethics for scientists: A companion for students*. Oxford: Wiley-Blackwell.

acknowledged in the text, just not as authors. This is commonly done in an acknowledgements section at the end of a paper.

If there are doubts about authorship, these need to be discussed before any research is undertaken. Otherwise, it may be too late to correct any disputes when it is time to submit the manuscript. It is much easier to rearrange roles as first author and authorship order before beginning than midway through research. This requires open, honest communication and an understanding of authorship guidelines by all contributors. Researchers should be wary of the pitfalls of collaborations involving multiple groups or universities, as there may be separate best practices or usual authorship expectations. If multiple parts of the paper have been divided among researchers, such that different paragraphs or different sections were independently and jointly authored by these individuals, then usually a footnote in the text explains that the work was divided and to whom.

Overall, regardless of the division of work in a paper, all authors testify to the accountability and truthfulness of their manuscript. If a journal article is published, all the authors bear the responsibility if misconduct regarding the article is discovered. Even if one author on the team fabricated data without the others' knowledge, the other authors could still be held liable for failing to detect such misconduct. This joint liability prevents authors from simply blaming a single scapegoat should a mutual fraud be uncovered. For instance, in the Korean stem-cell professor incident, the professor himself admitted that he had failed to exercise proper oversight over his lab and ordered his lab workers to fabricate data; at the same time, the investigation revealed that one co-author had also switched stem cells without telling others.[120]

Moreover, it can be extremely difficult for an author whose name is included on a fraudulent paper to remove their name and disassociate themselves from misconduct. Returning to the example of the Korean professor, when his American co-author wanted to remove his name from this Korean professor's 2005 manuscript after revelations of misconduct in the lab, *Science* denied his request.[121] Because the American co-author had agreed to a statement in the submission process that all authors take responsibility for the integrity of the research, *Science* decided that it was far too late to suddenly resign as an author despite the merits of the claims of misconduct on the part of the Korean professor.[122] Verification of the integrity of a published article should come before the submission process.

In a similar case, surrounding the cloned

---

[120] Cyranoski, D. (2014). Cloning comeback. *Nature*, 505(7484), 468-471. doi:10.1038/505468a

[121] Stewart, C. N. (2011). *Research ethics for scientists: A companion for students*. Oxford: Wiley-Blackwell.

[122] Stewart, C. N. (2011). *Research ethics for scientists: A companion for students*. Oxford: Wiley-Blackwell.

sheep Dolly, a legal hearing raised doubts about credit for the breakthrough. Initially, one person had been credited with the creation of Dolly through SCNT-cloning.[123] After a legal case, this person admitted that another collaborator actually deserved 66% of the credit for the work and was not lead author due to a prior agreement.[124] Additionally, two technicians who had done substantial amount of work were denied co-author status due to their positions in the laboratory, instead being relegated to an acknowledgements section at the end of the publication.[125]

In the worst-case scenario, authors who are unable to agree on a specific order listing their respective contributions may find themselves retracting their paper. Such was the case when Chinese authors who had unsolvable difficulties in finding a proper order for co-authors removed two

---

[123] Savulescu, J. (2008, March 01). Research ethics and lessons from Hwanggate: What can we learn from the Korean cloning fraud? Retrieved from https://jme.bmj.com/content/34/3/214?legid=medethics;34/3/214

[124] Savulescu, J. (2008, March 01). Research ethics and lessons from Hwanggate: What can we learn from the Korean cloning fraud? Retrieved from https://jme.bmj.com/content/34/3/214?legid=medethics;34/3/214

[125] Savulescu, J. (2008, March 01). Research ethics and lessons from Hwanggate: What can we learn from the Korean cloning fraud? Retrieved from https://jme.bmj.com/content/34/3/214?legid=medethics;34/3/214

papers from *Molecular Medicine Reports*.[126] Authors should openly discuss questions about authorship before a study begins, as well as during the course of research.

Even worse, omitting other personnel who did work for the material in a manuscript can result in a retraction and replacement. This was recently addressed when an optics journal pulled a 2016 manuscript after learning that the sole author at a British university had listed himself as the only author. The author had stolen credit for the collective work of a group and team.[127]

The opposite scenario, adding fictitious co-authors, is also not permitted. A physician pretended to be a biomedical researcher in over 60 papers on cancer research by often including non-existent, fictitious, co-authors, when he was actually just a medical practitioner not involved with any research team.[128]

---

[126] Marcus, A. (2018, September 27). "Irreconcilable" differences about author order, other issues topple two articles in Spandidos journal. Retrieved from https://retractionwatch.com/2018/09/27/irreconcilable-differences-about-author-order-other-issues-topple-two-articles-in-spandidos-journal/

[127] Stern, V. (2018, April 09). Journal retracts and replaces paper because author stole credit for group's work. Retrieved from https://retractionwatch.com/2018/04/10/journal-retracts-and-replaces-paper-because-author-stole-credit-for-groups-work/#more-63965

[128] Broad, W. J. (1980). Would-Be Academician Pirates

Adding co-authors who have not substantially contributed to a publication is known as honorary authorship, guest authorship, or gift authorship. These are all violations of academic integrity, where a politically powerful, wealthy, or senior member of a research organization is granted the honor of joining their name to a manuscript though they are not a part of the research. Typically, this is a hierarchical action demanded by those who supervise laboratories or manage multiple groups of mentored students. It is an unfortunately common ethical breach and is pervasive in East Asian research. However, this trend is slowly changing and more researchers are becoming aware of the need to deny these requests or demands for favors by those who rank above them in age or social hierarchies. In addition, an emphasis on the quality of publications by a researcher, rather than the number of publications, will help remedy this issue. Recently, allegations surfaced that professors at a famous university in Korea had listed their own children as co-authors on publications in order to improve their chances of being accepted when applying to college.[129]

Non-writing authors whose names do not appear on a manuscript may be part of a secret

---

Papers. *Science, 208*(4451), 1438-1440. doi:10.1126/science.208.4451.1438

[129] Zastrow, M. (2018). Kid co-authors in South Korea spur government probe. *Nature, 554*(7691), 154-155. doi:10.1038/d41586-018-01512-5

agreement who has written the paper for another. This is known as deception by ghost authorship, and private companies have engaged in using academic research to promote their products or discredit research that harms their ability to make money. For instance, agricultural biotechnology firm Monsanto had documents revealed during litigation that showed how Monsanto scientists designed deceptive academic papers, reviewed manuscripts, and even wrote and edited academic papers.[130] These papers authored by Monsanto attempted to discredit research that was critical of Monsanto products.[131] None of the papers included information about the Monsanto scientists' significant influence on the manuscripts.

The inverse also hold true. If for example, a university professor is collaborating with a collaborator at a biotechnology company, they may share data on a project and plan to act as co-authors. If, however, the professor wants to use that data for a separate grant proposal that the company is not a part of, there is an ethical issue with using the company's data. Because the

---

[130] Gillam, C. (2018, September 26). Carey Gillam and Nathan Donley: A story behind the Monsanto cancer trial - journal sits on retraction. Retrieved from https://www.ehn.org/monsanto-science-ghostwriting-2597869694.html

[131] Gillam, C. (2018, September 26). Carey Gillam and Nathan Donley: A story behind the Monsanto cancer trial - journal sits on retraction. Retrieved from https://www.ehn.org/monsanto-science-ghostwriting-2597869694.html

company does not want to participate in the grant proposal, they should not be listed on the grant, and neither should the professor take their data and use it for the proposal. Authorship without contribution, and contribution without authorship, carry significant ethical risks for research. With an increasingly large percentage of money for clinical drug trials in the United States and Korea coming from private industry, it is becoming more important to safeguard the rights of researchers to publish findings. Some researchers have ensured the freedom to publish all findings with the addition of publication clauses to ensure that all results can be published, including funding work that is disliked by or disagreeable to the publisher.

Seoul National University Research Ethics Guidelines

Revised on July 16, 2010.

Chapter 1 Responsibility and Obligation of Researchers

Article 2

【Responsibility as a Leader】

① A professor shall faithfully lead researchers and students so that their research may be performed creatively.

② A professor shall have his/her researchers and students become well-acquainted with ethics standard including research ethics guidelines that they should abide by in the course of performing research.

③ A professor shall impartially grant

authorship and distribute resources to researchers and students and treat them fairly.

④ A professor shall support future academic generations such as post-doctoral researchers, etc., so that they may grow into independent researchers.

Seoul National University Research Ethics Guidelines

Revised on July 16, 2010.

Chapter 2 Integrity in Research Presentation

Article 10

【Authorship of Work】

① In performing co-work, a researcher shall clarify the role of each researcher and mutual relations and do his/her best in order to fulfill his/her own responsibility.

② A researcher shall, without fail, indicate researchers who have made significant contributions to the planning of a research, establishment of concept thereof, performance, analysis, and preparation of performance result as authors or presenters.

③ When presenting a research result, the order of indicating authors or presenters shall be determined by the common consent of the participating researchers, and such decision shall be made considering the characteristics and reasonable practice of the relevant major field of study.

④ The institute/department where a presenter of a research result belongs shall be indicated as the institute/department where the presenter belonged at the time of performing experiments and writing, etc. However, where another practice is in common use in the relevant major field of study, such practice may apply.

⑤ A person who has not made a contribution to the planning of research, establishment of concept thereof, performance, analysis, and preparation of performance result shall not be included in the authors or presenters on grounds of acknowledging statement or honorable treatment, etc.

⑥ A principal investigator or a professor shall not urge researchers or students controlled or guided by himself/herself to give him/her authorship or billing that does not correspond with his/her contribution.

13. Data Management

Despite the increasingly large sets of data researchers use, beginning researchers often receive almost little to no formal training when it comes to recording, storing, and sharing data. Scientific data includes protocols, setup-up data, raw data, processed data, calculations, images, electronic files, lab notebooks, and the results from the experiment in various forms. When keeping laboratory notebooks, researchers should specify the time and location of the observations, including numerical and qualitative results. All data should be clearly labeled in order to serve as a record and resource to resolve any ambiguity even years afterwards should a conflict in data authenticity arise. This is critical when working in a laboratory, because all co-authors are responsible for attesting to the truthfulness of a published article. The NIH requires laboratory notebooks and all notebook to be retained for at least 7 years after creation.

Data can lead to a rejection of the study hypothesis and result in negative findings that are difficult to successfully submit for publication. As such, modifying the method of analysis to achieve desired results in an unethical violation, as is omitting data that does not support desired conclusions. Of course, these are also followed by fabrication and falsification of data in the worst instances of misconduct.

In other cases, data can be segmented and omitted to achieve desired results or multiple publications. Some medical journal editors may also refer to this practice as "salami" slicing, in reference to a type of thinly sliced meat similar to that used for shabu shabu. The metaphor describes how the overall research results are segmented into smaller, less meaningful pieces that are quite different from the result overall. For example, an article in *The Journal of the American Medical Association* on the drug Celebrex was discredited after it was revealed that the authors had only submitted 6 months' worth of data on the experiment when in reality 12 months' worth of data had been collected.[132] In other cases,

---

[132] Savulescu, J. (2008, March 01). Research ethics and lessons from Hwanggate: What can we learn from the Korean cloning fraud? Retrieved from https://jme.bmj.com/content/34/3/214?legid=medethi

authorship disputes may result in the creation of multiple, redundant papers in order to allow for different first authors. As such, disclosure of the process involved in collecting data can alleviate concerns about redundant or incomplete data in publications.

Seoul National University's Lab Note Guidelines

Revised on May 30, 2014. Established on January 1, 2015.

Chapter 3 Management of Lab Note

Article 9

【Preservation and Management】

① Lab notes shall be preserved according to the following subparagraphs:

1. In principle, a lab note shall be preserved for thirty (30) years from the date of preparation;

2. In case of retirement, temporary retirement, or change in participation, etc., a researcher shall hand over the lab notes prepared up to that time to the succeeding researcher or return them to the principal investigator.

② The head of the R&DB Foundation may preserve written lab notes using separate preservation methods such as documents subject to computerization, micro film, or other electronic media, etc.

③ The head of the R&DB Foundation shall

cs;34/3/214

prepare a management ledger of lab notes and record and maintain matters necessary for the management of lab notes.
Seoul National University Research Ethics Guidelines Revised on July 16, 2010. Chapter 3 Management of Research Material Article 14 【Principle】 Research data (referring to source materials such as material / process / result of experiment, observation / field investigation / survey result, etc.) and research materials (referring to research data and secondary material obtained from processing the research data) shall be clearly recorded as prescribed below so that other researchers may reproduce the same result under the same conditions. However, exception may be acknowledged according to the characteristics of a major field.
Seoul National University Research Ethics Guidelines Revised on July 16, 2010. Chapter 3 Management of Research Material Article 15 【Recording of Research Data】 ① Research data, as grounds certifying the

originality of research and which are able to give legal protection to a research result shall be correctly recorded in a lab note. ② Since a lab note, as an aggregate of all information occurring in the laboratory, is an important tool for handing down knowledge in the laboratory and to be used by succeeding researchers as well as the researcher himself/herself, it shall be correctly prepared in accordance with the following guidelines:1. A lab note, in principle, shall be prepared in writing using a durable writing tool and a bound note for the preservation of research data. It may be prepared in the form of electronic documents according to the characteristics of research. In such case, record certification function corresponding to a written lab note shall be supported; and2. Research data shall be recorded together with the research date and serial number in order to be protected from counterfeiting or forgery. A correction mark shall be made in cases wherein the recorded contents are amended so that the original contents may be discerned. Research data such as a printout from an experiment instrument, etc., and which cannot be written down in a lab note shall be pasted to the note or kept in a separate safe place, with the place of safe-keeping recorded in the lab note.
Seoul National University Research Ethics Guidelines Revised on July 16, 2010.

Chapter 3 Management of Research Material

Article 16

【Preservation of Research Material and Lab Note】

① The principal investigator shall assume responsibility for keeping and managing research materials and lab notes.

② Lab notes shall belong to the property of Seoul National University.

③ Research materials shall be preserved for the period deemed necessary for verification by the academic field (at least for five years).

④ Research materials requiring security shall be kept at a place not accessible to the general public, with an access password set up for computer files.

⑤ Any person who intentionally falsifies or destroys research materials or research data shall bear the consequence as prescribed by Article 7, Chapter 2.

Seoul National University's Lab Note Guidelines

Revised on May 30, 2014. Established on January 1, 2015.

Chapter 2 Preparation of Lab Note

Article 5

【Requirements of Lab Notes】

① A written lab note shall meet the requirements falling under the following subparagraphs:

1. As for a lab note, a bound form of note wherein insertion and deletion is difficult shall be used;

2. When writing contents, a writing tool that makes the contents unchangeable in quality and preservable for a long time shall be used;

3. On the first page of a lab note, the name of the institution, title of project, name of principal investigator, name of the recorder, and lab note management number shall be written down. On each page, its own page number shall be printed, and an edge with a margin shall be kept. In addition, the signatures of a recorder and an inspector, date of recording, date of inspection, etc., shall be included in each page.

② An electronic lab note shall meet the requirements under the following subparagraphs:

1. The function of certifying the recorder's signature on electronic documents shall be included;

2. The function of automatically recording the date and time research records are inputted shall be included; and

3. The function of identifying forgery or alteration made in the records shall be included.

14. Publication

Publication is the means by which the scientific community publicizes findings by researchers. Along with authorship, it is part of the process by which credit and achievement is validated for scientists. Academic journals are one of the primary ways that the process of discussing and evaluating the findings from research occurs. Additionally, publication has a role in the allocation of funding research by government grants and other investors in research. Some countries also offer cash prizes to professors who can publish articles in high profile, prestigious journals.

These factors can result in an intense pressure for academics and researchers to publish a large quantity of papers, valuing speed and output over meaningful scientific discovery, quality research, and collaboration. This is a particular issue in Korea for many researchers, when researchers lack the ability to critically evaluate the actual usefulness and contribution of research. Instead, exaggerated titles of papers that bear little connection to their content and a race to publish many low-quality papers can tarnish the reputation of the Korean scientific community.[133] Additionally, the name of the journal and the ranking of the journal is often a substitute for the actual content of the manuscripts that are published within the journal.[134]

Finally, the increasingly interconnected nature of globalization and the internet mean bad academic research and poor publication practices can quickly reach wide audiences and contaminate large sections of research. It can be very difficult for research papers that are retracted from academic journals to be removed from every paper that has used a retracted paper for a citation. Because scientific publication serves the goal of the reciprocal relationship between researchers and

---

[133] Nho, H. (2016). Research ethics education in Korea for overcoming culture and value system differences. *Journal of Open Innovation: Technology, Market, and Complexity, 2*(1). doi:10.1186/s40852-016-0030-3

[134] Savulescu, J. (2008, March 01). Research ethics and lessons from Hwanggate: What can we learn from the Korean cloning fraud? Retrieved from https://jme.bmj.com/content/34/3/214?legid=medethics;34/3/214

the public in communicating their breakthroughs and discoveries to the public and society, it is unethical to use publications for selfish, unethical goals such as artificially inflating a researcher's record of publications. Moreover, because medicine has a particularly important impact on human health and patient safety, clinical studies involving misconduct or data issues can place human subjects at a risk of harm. To be clear, publications benefit society, while ensuring that the scientific process allows for verification, replication, reproduction, and analysis of each paper in conjunction with the incremental advances that allow science to improve and mature.

> Seoul National University Research Ethics Guidelines
>
> Revised on July 16, 2010.
>
> Chapter 2 Integrity in Research Presentation
>
> Article 5
>
> 【Principle】
>
> A researcher shall, in presenting his/her research results, maintain academic integrity as much as possible by complying with the following matters:
>
> 1. He/She shall make a correct, truthful description when reporting or presenting his/her own research results;
>
> 2. In cases wherein a researcher finds a matter that violates these research ethics guidelines after presenting his/her research

results, he/she shall withdraw without delay all or part of the research results; and 3. He/She shall not publish his/her own research result in mass media in an exaggerated manner.
Seoul National University Research Ethics Guidelines Revised on July 16, 2010. Chapter 2 Integrity in Research Presentation Article 9 【Limitation on Multiple Publications】 ① A researcher shall not publish multiple times all or part of his/her own theses or writings that have already been published in the same language or other languages without correct indication of its source and citation. Where research data or sentences, even if partially different from each other, are recognized as identical with each other as a whole, the same shall apply. ② Notwithstanding the provision of Paragraph (1), a researcher may make a publication falling under any of the following subparagraphs (in cases of Subparagraphs (1) ~ (6), however, correct indication of such source or citation shall be made; however, an exception may be made considering the characteristics of a major field of study and the opinions of the relevant academic field): 1. When publishing all or part of his/her

thesis in a separate thesis or book;

2. When publishing all or part of his/her research result report in a separate thesis or book;

3. When gathering together already published theses in a book;

4. When publishing all or part of the same thesis or book in the same or different language with approval from the copyright owner;

5. When, in publishing such thesis after a brief paper (letter, brief communication, etc.) has been published in an academic journal and changed into a long thesis, the researcher publishes the brief letter paper by adding research data, interpretation, or information, etc., in the course of performing research thereto;

6. When all or part of a thesis or a book already published is selected, compiled, and published in the form of anthology by another editor or published in a special issue of an academic journal;

7. When all or part of a thesis or a book already published is written in an easy-to-understand manner and published in a non-academic publication such as educational book, popular magazine, etc.; or

8. Other publication corresponding to any of the above subparagraphs and which does not affect academic integrity.

③ The registration of research results that have already been presented into intellectual property rights shall be permitted, notwithstanding Paragraphs (1) and (2.)

15. Peer Review

Peer review is part of the process that science steers the cultural direction of progress. That is to say, peer review is the steering wheel behind the ship of publication. Every reputable scientific journal requires potential publications to undergo peer review in the submission process.

After submitting an unpublished article to a journal, the article is reviewed initially by the journal editors. This first line of review results in a substantially high rate of rejection for potential manuscripts. If a publication manages to survive this review, the chances for publication increase, but it must then be approved by peer reviewers. What is a peer reviewer? A peer reviewer is a peer or fellow member of the research community to which the article is being published. Usually, peer reviewers are other editors are authors who submit to the journal or related research areas. They are experts in these fields and as such can detect issues with experimental design, methods, and overall questions that readers may have if the manuscript is published. Peer reviewers hold the power to accept, reject, or recommend further changes pending resubmission to a manuscript. Because these are unpublished manuscripts, they may contain novel ideas and findings that no one besides the author and the peer reviewer are aware of. This places the peer reviewer in a position of confidentiality and an ethical obligation not to disclose the unpublished, novel content or important information from the peer reviewed article to anyone else. Similarly, they may not use any information that is newly found in the manuscript for themselves.

Peer reviewers are also used in government grant and funding applications. In those scenarios, the same rules and guidelines apply to not disclosing privileged information from the grant submission outside of the peer reviewer role.

In the realm of journals, an investigation by the NIH found that a researcher stole ideas from the unpublished manuscript of a rival researcher and used that information to develop a study and later published in the famous journal *Science*.[135]

---

[135] Specter, M. (1989, June 13). NIH Accuses Biologist of Stealing Ideas From Rival Researcher. *Washington Post*.

The researcher had been given ideas for experiments and information from a manuscript submitted by an Ivy League professor. Instead of evaluating the manuscript, the researcher used it for his own professional attempts at achieving success by failing to acknowledge the manuscript's role in his later publication, as well as abusing his role as peer reviewer to misappropriate privileged information. This is a violation of the peer review process, and harms the implicit integrity of those who serve as peer reviewers. These types of peer review incidents can exacerbate the worrying trend of growing mistrust in science.[136]

Korean science is not immune to peer review issues. Another former Korean professor used fake names and email addresses to fool journal editors into thinking his work had been reviewed.[137] Some trusting journals allow submission to include potential contacts suggested

---

Retrieved from https://www.washingtonpost.com/archive/politics/1989/07/13/nih-accuses-biologist-of-stealing-ideas-from-rival-researcher/fd8a3122-c461-49d8-bd05-c8b99e930f2d/

[136] Check, E., & Cyranoski, D. (2005). Korean scandal will have global fallout. *Nature, 438*(7071), 1056-1057. doi:10.1038/4381056a

[137] Oransky, I. (2014, November 10). Retraction count grows to 35 for scientist who faked emails to do his own peer review. Retrieved from https://retractionwatch.com/2012/09/17/retraction-count-for-scientist-who-faked-emails-to-do-his-own-peer-review-grows-to-35/

for peer reviewer who may be familiar with background information on the research. The professor took advantage of this trust and then submitted his own falsified, overly positive peer reviews to make it seem like others had reviewed his submissions. How was he caught? The professor's papers created suspicion when it was discovered that all his reviews came back shockingly fast, within 24 hours. A normal review takes time to carefully evaluate and comment on a prospective manuscript, and it would be highly unusual for all to come back so quickly. Since discovery of the professor's fraud in 2012, he has had 30 papers retracted, including 4 retractions alone from the *Journal of Enzyme Inhibition and Medicinal Chemistry*. Presumably, the professor was not interested in advancing scientific progress and more fixated on making money quickly by artificially inflating his CV and career.[138]

---

[138] Oransky, I. (2014, November 10). Retraction count grows to 35 for scientist who faked emails to do his own peer review. Retrieved from https://retractionwatch.com/2012/09/17/retraction-count-for-scientist-who-faked-emails-to-do-his-own-peer-review-grows-to-35/

16. Reproducibility

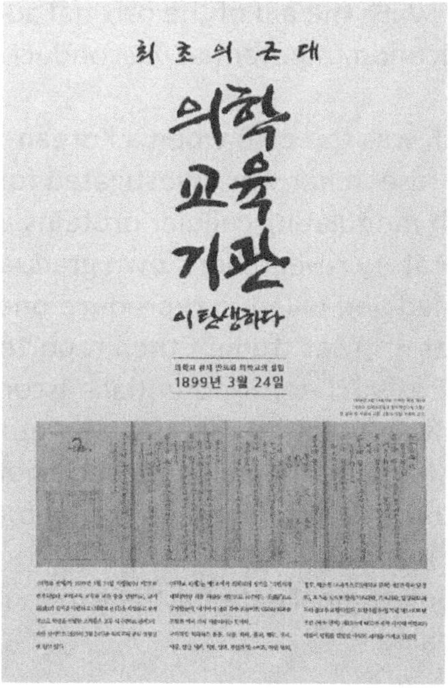

Publishing a scientific manuscript includes a methods and materials section that provides all of the information for independent researches to perform their own successful re-enactment of the experiment. This ability for publications to be redone or reattempted by others is known as reproducible or replicable research.

Reproducible research means research findings that can be verified by others, and is a sign of a quality contribution to academic innovation and progress. In contrast, a study that fails to replicate can sometimes be a sign of underlying issues with the way the experiment was designed. In the worst cases, research findings that fail to replicate even after many attempts by different researchers with the aid of the original author can be an implication of potential misconduct.

Such was the case when a Korean laboratory researcher was investigated for falsified research on modulating cellular proteins after one of the laboratory researcher's own graduate students tried and failed to reproduce one of his experiments.[139] That student then reported the laboratory researcher for potential misconduct, and the university found that the laboratory researcher had intentionally changed photos in order to fabricate results. When the laboratory researcher was unable to locate laboratory notebooks or original data for the experiments, his fate was sealed. [140]

More often, however, reproducibility may be undertaken by those at other institutions. When a team consisting of Harvard, Cornell, Cambridge, and other university researchers could not reproduce findings from their own published paper

---

[139] Korean researcher fired for fraud. Retrieved from https://www.the-scientist.com/the-nutshell/korean-researcher-fired-for-fraud-45361

[140] Science. (2017, December 10). South Korean Researcher Suspended Over Charges of Scientific Misconduct. Retrieved from https://www.sciencemag.org/news/2008/03/south-korean-researcher-suspended-over-charges-scientific-misconduct

from 2013 in a *Cell* journal, they decided to retract their own manuscript because of the doubts that a lack of reproducibility caused with regard to the validity of their findings.[141] Behaving ethically, they erred on the side of caution in deciding that such results needed to be reproducible to verify their findings. This was a positive, pre-emptive way of anticipating irreproducible findings. Reproducibility is a tool for evaluating results by others and for the authors themselves.

Currently, biomedical research is in crisis worldwide. As much as 90 percent of the published medical information that physicians rely on is flawed.[142] For most study designs, research claims are often more likely to be false than true.[143] Pervasive issues stemming from bias, conflicts of interest, and other medical research issues may impose chronic resistance to change or even disclosure of flawed research methodologies.[144]

---

[141] Abritis, A. (2018, May 18). Caught Our Notice: A team from Harvard, Cornell, Cambridge, HHMI, and UCSF can't reproduce a paper's findings. Retrieved from https://retractionwatch.com/2018/05/21/caught-our-notice-a-team-from-harvard-cornell-cambridge-hhmi-and-ucsf-cant-reproduce-a-papers-findings/#more-65829

[142] Freedman, D. H. (2015, September 02). Lies, Damned Lies, and Medical Science. Retrieved from https://www.theatlantic.com/magazine/archive/2010/11/lies-damned-lies-and-medical-science/308269/

[143] Ioannidis, J. P. (2005). Why Most Published Research Findings Are False. *PLoS Medicine, 2*(8). doi:10.1371/journal.pmed.0020124

Almost every step of the scientific process can be easy manipulated, even unintentionally or unconsciously.[145] It has been reported that a research finding can be less likely to be true where effect sizes are smaller, more relationships are tested, more research teams involved, and greater flexibility is allowed during research as well as where various interests and prejudice are involved.[146]

---

[144] Freedman, D. H. (2015, September 02). Lies, Damned Lies, and Medical Science. Retrieved from https://www.theatlantic.com/magazine/archive/2010/11/lies-damned-lies-and-medical-science/308269/

[145] Freedman, D. H. (2015, September 02). Lies, Damned Lies, and Medical Science. Retrieved from https://www.theatlantic.com/magazine/archive/2010/11/lies-damned-lies-and-medical-science/308269/ ("There was plenty of published research, but much of it was remarkably unscientific, based largely on observations of a small number of cases." "[Ioannidis] discovered that the range of errors being committed was astonishing: from what questions researchers posed, to how they set up the studies, to which patients they recruited for the studies, to which measurements they took, to how they analyzed the data, to how they presented their results, to how particular studies came to be published in medical journals." ("At every step in the process, there is room to distort results, a way to make a stronger claim or to select what is going to be concluded," says Ioannidis. "There is an intellectual conflict of interest that pressures researchers to find whatever it is that is most likely to get them funded.")

[146] Ioannidis, J. P. (2005). Why Most Published Research Findings Are False. *PLoS Medicine*, 2(8). doi:10.1371/journal.pmed.0020124 (that is because the probability that a research claim is true may depend on

As a result, initially promising findings across biomedical research failed to lead to improvements in health care.[147] The statistical significance highlighted in almost all articles on cancer prognostic marker studies loses its discriminating ability for the importance of prognostic markers under strong reporting bias.[148] Although more than 1,500 reports detailed at least one significant cancer prognostic variable in 2005,[149] few identified biomarkers have been confirmed by subsequent research and few have entered routine clinical practice.[150] Consequently, it has been

---

study power and bias, the number of other studies on the same question, and, importantly, the ratio of true to no relationships among the relationships probed in each scientific field) ("Moreover, for many current scientific fields, claimed research findings may often be simply accurate measures of the prevailing bias.")

[147] Macleod, M. R., Michie, S., Roberts, I., Dirnagl, U., Chalmers, I., Ioannidis, J. P., Glasziou, P. (2014). Biomedical research: Increasing value, reducing waste. *The Lancet, 383*(9912), 101-104. doi:10.1016/s0140-6736(13)62329-6

[148] Kyzas, P. A., Denaxa-Kyza, D., & Ioannidis, J. P. (2007). Almost all articles on cancer prognostic markers report statistically significant results. *European Journal of Cancer, 43*(17), 2559-2579. doi:10.1016/j.ejca.2007.08.030

[149] Kyzas, P. A., Denaxa-Kyza, D., & Ioannidis, J. P. (2007). Almost all articles on cancer prognostic markers report statistically significant results. *European Journal of Cancer, 43*(17), 2559-2579. doi:10.1016/j.ejca.2007.08.030

[150] Anderson, N. L. (2009). The Clinical Plasma Proteome: A Survey of Clinical Assays for Proteins in Plasma and

estimated that over 85 percent of biomedical research investment is cumulatively wasted, the equivalent of $200 billion globally as of 2010.[151][152]

The crisis in reproducibility has its roots in a variety of factors originating from the levels of researchers, research institutions, and industry. For biomedical research, there are proactive steps possible to help confront these challenges, such as funding decisions, research conduct, research regulation and management, research report, information accessibility, etc. Given the unique nature of universities in providing both education and research, they have a singular burden in resolving this crisis.[153] Particularly, universities can provide institutional oversights and guidelines for research integrity and reproducible science, compensating for any insufficiencies in external regulatory regimes.

Serum. *Clinical Chemistry, 56*(2), 177-185. doi:10.1373/clinchem.2009.126706

[151] Chalmers, I., & Glasziou, P. (2009). Avoidable waste in the production and reporting of research evidence. *The Lancet, 374*(9683), 86-89. doi:10.1016/s0140-6736(09)60329-9

[152] Macleod, M. R., Michie, S., Roberts, I., Dirnagl, U., Chalmers, I., Ioannidis, J. P., Glasziou, P. (2014). Biomedical research: Increasing value, reducing waste. *The Lancet, 383*(9912), 101-104. doi:10.1016/s0140-6736(13)62329-6. pmid:24411643

[153] Freedman, L. P., Cockburn, I. M., & Simcoe, T. S. (2015). The Economics of Reproducibility in Preclinical Research. *PLOS Biology, 13*(6). doi:10.1371/journal.pbio.1002165

Reproducible science matters to South Korea, which intends to become a true innovator with cutting edge research. South Korea has spent over 4% of its gross domestic product (GDP) for R&D to drive its economic growth. Since 2005, South Korea has more than doubled academic publication output in many scientific fields including life sciences and has many patent applications filed. South Korea should pay much more attention to reproducible science in order to transform into a world leader in basic science from an industry-focused "fast follower." Specifically, research universities in South Korea are expected to play a central role in promoting reproducible science.

17. Cognitive Bias

New studies regarding the psychology and science behind the human aspect of research mistakes have grown increasingly popular in the scientific community. Although this discussion of research ethics focuses on biomedical perspectives, beginning researchers should be prepared to encounter more frequent topics on cognitive bias, and the way it can influence the actions of researchers in unknown, unnoticed ways. Common dialogues on cognitive bias point to a subconscious desire for specific hypotheses to be proven true, which then influence data interpretation and study reporting. Such is the rationale for traditional scientific method protections such as data blinding.

More recently, practices such as blind analysis, or removing data labels on data when statistical, post-experimental analysis is conducted, have become more common. Some commentators have also referred to this in the context of "triple blinding" for research ethics. Moreover, sharing data is another way to mitigate the potential for accidental or inadvertent mental processes to influence a study's outcome. By posting data online, in specific repositories, or amongst research groups and colleagues, a researcher can ensure that additional viewpoints from the research community can guard against the potential influence of cognitive bias.

Finally, cognitive bias has an especially critical role in reproducible research. Although beyond the scope of this work, cognitive bias has been shown to play a vital role in establishing standards and practices for reproducible research. So far, many of the examples of research ethics mistakes have been open and obvious. However, as research ethics grows more sophisticated, more scrutiny is focused on unconscious cognitive biases. A new researcher will surely see more on this topic throughout the course of a biomedical career, and should start preparing mainstream adoption of data sharing and blind analysis as demand for reproducible research gains momentum.

18. Conflict of Interest

Traditional research ethics education would be imprudent without mentioning the concept of a conflict of interest. At the same time, it is worth noting the obvious fact that while the idea of a conflict of interest as commonly known in Western notions of academic and other fields is a relatively established ethical ideal, in East Asia these ideas may face greater barriers to acceptance due to longstanding cultural traditions. With this distinction in mind, teaching about the harms of conflict of interest is approached from a viewpoint that even though there may be cases where conflict of interest in unavoidable, there is a still a need for research ethics education to confront the pervasive issue and its effects.

In sum, there is often a double standard in research ethics when confronting conflicts of interest in Korea: while students may understand and accept the need to be aware of conflicts of interest, oftentimes advocating these lessons outside of the classroom can meet apathy from society at large.[154]

The idea of a conflict of interest refers to

---

[154] Nho, H. (2016). Research ethics education in Korea for overcoming culture and value system differences. *Journal of Open Innovation: Technology, Market, and Complexity, 2*(1). doi:10.1186/s40852-016-0030-3

those relationships and situations where the people involved have the potential to act in an unethical manner to benefit another person, organization, or idea unfairly. These conflicts are clashes between a person's self-interest, the interests of their friends or family, and are very broad in nature.

The previously discussed case of the Korean professor, the former stem-cell researcher, also illustrates issues with authorship, conflict of interest, and unethical behavior. This professor faced a unique individual role in Korea as someone who had access to the valuable personal networks and privileged connections to politicians, administrators, and other powerful people.[155] Because grants often come from government funding sources, researchers may feel compelled to base their research on political viewpoints. The professor added a government advisor to his 2014 retracted article as an honorary author—an ethical violation. The professor did this in gratitude for the generous financial support that he had been given.[156] However, this was also a potential conflict of interest that went undisclosed. Because the professor received funding provided by those who

---

155 Gottweis, H., & Triendl, R. (2006). South Korean policy failure and the Hwang debacle. *Nature Biotechnology, 24*(2), 141-143. doi:10.1038/nbt0206-141

156 Gottweis, H., & Triendl, R. (2006). South Korean policy failure and the Hwang debacle. *Nature Biotechnology, 24*(2), 141-143. doi:10.1038/nbt0206-141

were working directly with the government, it is foreseeable that the same paper could be used to obtain further funding from that same office. In essence, the government would be funding research for which it was a co-author and contributor with the professor.

A failure to disclose significant financial conflicts of interest can be seen as research misconduct if severe enough. Often, researchers may have financial interests in areas that are potentially affected by their research. This could take the form of pharmaceutical company consulting, funding from industry companies, biotechnology investments, and other such things. While an individual researcher may strongly believe that he or she is capable of resisting or succumbing to influence by means of conflicts of interests. However, in aggregate, these ties have a pervasive effect on the outcome of research and, as previously noted in cognitive bias descriptions, are often unconscious in nature. For example, research on the harmfulness of passive (second-hand) smoking for 106 articles published between 1980 and 1995 found that tobacco-affiliated researchers were significantly more responsible for articles concluding that smoking was not harmful.[157] Finally, public perception of the value and safety of clinical

---

[157] Savulescu, J. (2008, March 01). Research ethics and lessons from Hwanggate: What can we learn from the Korean cloning fraud? Retrieved from https://jme.bmj.com/content/34/3/214?legid=medethics;34/3/214

research is harmed by conflicts of interest that are later revealed rather than disclosed beforehand.[158] With increased public scrutiny due to the Improper Solicitation and Graft Act (Kim Young-ran Act) and public scandals involving researchers and payments, it is important for researchers to understand the trend that conflicts of interests epitomize.

Seoul National University Research Ethics Guidelines Revised on July 16, 2010. Chapter 1 Responsibility and Obligation of Researchers Article 3 【Responsibility When Receiving Research Expenses Support】 ① A researcher shall, if he/she receives full or partial financial support for the research expenses, properly disclose the fact and transparently manage the financial resources. ② In case of receiving support for research expenses, academic integrity and freedom shall not be restricted. ③ In cases wherein an academic essay is prepared based on the results of a research service or materials obtained in the course of research service, the fact of receiving support for the

[158] Botkin, J. R. (2018). Should Failure to Disclose Significant Financial Conflicts of Interest Be Considered Research Misconduct? *Jama, 320*(22), 2307. doi:10.1001/jama.2018.17525

research expenses, in principle, shall be specified.
Seoul National University Research Ethics Guidelines Revised on July 16, 2010. Chapter 4 Conflict of Interests Article 17 【Principle】 In cases wherein, regarding the planning of research, collection/analysis/interpretation of materials, publication, use of results, etc., conflict of interests prescribed by Article 18 occur -- or are likely to occur -- on the part of a researcher, a researcher shall manage such conflict of interests in accordance with Article 19 since fairness of research, safety of the human subject of research, and social trust in academic research might be impaired.

Seoul National University Research Ethics Guidelines Revised on July 16, 2010. Chapter 4 Conflict of Interests Article 18 【Contents of Conflict of Interests】 The term "conflict of interests" refers to cases that may have a negative effect on professional judgment or research performance due to any of the following causes: 1. Monetary conflict of interests: when caused by pecuniary advantage of a researcher related to research; 2. Conflict of interests in human relations: when caused by personal human relations such as personal acquaintance, influence by the institution where he/she belongs, personal conflict or research competition, etc.; 3. Intellectual conflict of interests: when caused by religious belief, world view, or moral conviction or theoretical conviction with respect to a certain kind or field of research; 4. Conflict of interests due to role conflicts: when roles such as education, services, and external activities as a member of the research institution where he/she belongs collide with research activities; or 5. Other conflict of interests: when caused by other reasons corresponding to Subparagraphs 1 ~ 4.
Seoul National University Research Ethics Guidelines

Revised on July 16, 2010.
Chapter 4 Conflict of Interests
Article 19

【Management of Conflict of Interests】

① In cases wherein there is any actual or potential possibility of conflict of interests relating to research arising, a researcher shall do his/her best so that such conflict may not have a negative effect on research by ensuring transparency and accountability by disclosing such conflict to the public.

② A researcher shall specify in a research proposal matters such as support details of research expenses, supporters, institution where he/she belongs, and other grounds for causing conflict of interests under Article 18 and disclose such facts in an academic journal even when presenting research results. Furthermore, if necessary, the researcher shall notify the human subject of research thereof.

③ In cases wherein there is a possibility of a negative effect on research actually arising since the degree of conflict of interests is serious, a researcher shall suspend research without any delay or, if continuing the research, shall request supervision over the fairness of research to an expert group independent of such conflict of interests. To ensure that the fairness of research is not impaired by a co-worker who has conflict of interest, the principal investigator shall take an appropriate measure such as exclusion of such co-worker from a specific research stage.

Hypothetical:

Prof. Z is a professor at a Korean medical school. His academic focus is on genomic research and neuroscience, and he serves as a peer reviewer for several academic journals in this area. His older sister, Dr. Z, is also involved in scientific research, as a business executive at the Korea Kimchi Fantastic Food Fund (KKFFF). KKFFF is attempting to use genetic modification and CRISPR-based technology to engineer varieties of vegetables that provide memory-boosting effects when processed for use in kimchi or other foods and then eaten by human consumers. In particular, KKFFF is also considering developing a line of dried vegetable snacks for university students who do not have time to eat proper meals when studying diligently for their exams. Prof. Z speaks with his sister regularly about the business and offers suggestions at times, but he is not employed by KKFFF. His sister holds large amounts of stock in KKFFF, and is considering partnering with Conglomerate Cooking Company (CCC) in order to expand their product lines. One day, Prof. Z receives an email from the *Journal of Korean Vegetable Genetic Research and Neuroscience*. The email asks him to serve as a peer reviewer for an article submitted to the journal on vegetable genetic-modification and its effects on intelligence. The article could have severe consequences for Prof. Z's sister, because its findings suggest that his company's product line will probably not have any intended effects on human memory when eaten, and this may even cause the potential partnership with CCC to fail. Still, Prof. Z believes the article may not quite be ready for publication due to some questions regarding their methodology and statistical analysis. However, it is possible that the journal could accept the manuscript after suggested changes are made. Prof. Z is unsure whether there is a conflict of interest. What should Prof. Z do? May he contact his sister about the article? May he submit his peer review evaluation to the journal?

19. Human Subjects and Informed Consent

Clinical research ethics is comprised of the ethical principles to protect human subjects in new treatment development, diagnosis, and prevention for patients through clinical trials. Here, research is defined as "a systematic investigation designed to develop or contribute to generalizable knowledge" under the Common Rule (U.S. Department of Health and Human Services Title 45 CFR 46). Alternatively, research is "any experiment that involves a test article, and one or more human subjects; and that is subject to regulatory agency such as FDA oversight or whose results are intended to be submitted to regulatory agency as part of an application for a research or marketing permit." The term human subject means "a living individual about whom an investigator conducting a research obtains data through intervention or interaction with the individual, or identifiable private information." Clinical research employing human subjects requires IRB review before starting the research.

Clinical research must be differentiated from medical care. Clinical research and medical care have different purposes, methodologies, and justification for the risks. The goal of research is not to help particular patient or population. Clinical research intends to answer scientific questions in order to benefit to future patients while medical care intends to provide personalized care here and now. Clinical research often employs randomization, blinding, placebo, and tests for research; medical care provides individualized diagnostic and treatment choices. The risks accompanying clinical research can be justified by the value of knowledge to be gained by clinical trial and by societal benefit; those accompanying medical care can be compensated by direct medical benefit to particular patient.

Regulation of clinical research has been swinging from no regulation to the establishment of specific laws and regulations for human subjects. The attitude toward clinical research has also changed from focusing on harm to focusing on the benefits from the research. The regulatory regime for human subject protection consists of codes, declaration, reports, and guidelines, as well as laws and regulations. The regime has been enforced by making human subject protection as a requirement for research funding by government, for research approval at an institution, for publication, and for marketing approval by regulatory agency.

According to the U.S. NIH, the most important ethical principles are: collaborative partnership, valuable scientific question, valid scientific methodology, fair subject selection, favorable risk-benefit ratio, independent review, informed consent, and respect for enrolled subject.

In one hand, many trials are failed because they cannot recruit a sufficient number of participants, on the other hand, some ethical issues have not been resolved yet. Whether a placebo-controlled trial is ethical and whether the researcher is obligated to provide treatment to participants after clinical trials are the examples. Research studies using human specimens, data, and information have raised specific ethical questions regarding consent for future research, recontact, the balance between broad access to datasets and privacy, sharing research results and incidental/secondary results with participants, sharing financial interests with participants, etc. Particularly, new DNA technology, and data science have exposed participants to the informational risks such as re-identification and data security in addition to traditional physical, social/psychological, and financial risks.

When dealing with human subjects, an

ethical researcher must make sure to adhere to the highest principles of respect for persons. Part of this involves receiving informed consent from each and every participant who is a subject to research. What is informed consent? Informed consent is a subset of human subject respect. It involves the process, from start to finish, of securing the understanding, affirmations, and consent to participate in clinical research and the way the data collected will be used. It is not simply the signing of a piece of paper, but rather involves continued respect for human subjects throughout their involvement.

However, the consent must be voluntary. It cannot involve coercion or an abuse of power, such as *gapjil* and associated hierarchical exercises of power. This is illustrated with reference again to the former stem-cell researcher and Korean professor. The professor's media coverage also raised questions about another incident, involving those working in his laboratory. The professor initially had claimed that women who had donated eggs for the stem-cell research had done so voluntarily, however, it later came out that donors had been paid significant sums of money. In addition, many donors also failed to receive adequate information about risks from the procedure to harvest eggs, and 16 women were hospitalized from adverse side effects.[159] Moreover, a report by *Nature* in May 2005 definitively

---

[159] Cyranoski, D. (2006). Rise and fall. *News@nature*. doi:10.1038/news060109-8

provided evidence that the professor's junior team members and laboratory researchers had been egg donors.[160] Many believed that it was a conflict of interest for the professor to use his own team members to contribute to the research in this manner, not to mention the painful and risky nature of the donation procedure.[161] Indeed, the lack of accountability for the professor surrounding the egg donations raised concerns for many. Critics noted a growing concern about whether the laboratory workers would actually donate their eggs so easily.[162] This was echoed by an anonymous website, which suggested that the junior team members had been coerced into donating their eggs.[163] Later, the National Bioethics Committee repeated this notion. After hundreds of eggs were ruined in early experiments due to a mistake, it was suggested that a research fellow in the laboratory she might have been pressured into donating.[164] Finally, the professor admitted he had

---

[160] Cyranoski, D. (2006). Rise and fall. *News@nature*. doi:10.1038/news060109-8

[161] Cyranoski, D. (2006). Rise and fall. *News@nature*. doi:10.1038/news060109-8

[162] Cyranoski, D. (2004). Koreas stem-cell stars dogged by suspicion of ethical breach. *Nature, 429*(6987), 3-3. doi:10.1038/429003a

[163] Savulescu, J. (2008, March 01). Research ethics and lessons from Hwanggate: What can we learn from the Korean cloning fraud? Retrieved from https://jme.bmj.com/content/34/3/214?legid=medethics;34/3/214

[164] Cyranoski, D. (2006). Whos who. *News@nature*. doi:10.1038/news060109-9

lied: not only had two students donated eggs for the study, the professor transported a student to the donation clinic himself.[165] Later, this student returned to the lab to attempt to make cell-line clones of her own biological material.[166] Specifically, this power disparity in the relationships with students who worked under him in his laboratory when handing out consent forms was an issue.[167] At the same time, this notion is controversial, inasmuch as it has been suggested that international guidelines such as the Declaration of Helsinki may face challenging in applicability to rapidly evolving areas such as stem-cells and egg donors.[168] It is arguable the extent to which a junior team member may have felt pressure to donate, especially given the equivocating in storytelling and initial lies by the professor.

The takeaway to all this is that when

---

[165] Cyranoski, D. (2014). Cloning comeback. *Nature, 505*(7484), 468-471. https://www.nature.com/news/cloning-comeback-1.14504#/timeline

[166] Cyranoski, D. (2014). Cloning comeback. *Nature, 505*(7484), 468-471. https://www.nature.com/news/cloning-comeback-1.14504#/timeline

[167] Cyranoski, D. (2014). Cloning comeback. *Nature, 505*(7484), 468-471. https://www.nature.com/news/cloning-comeback-1.14504#/timeline

[168] Savulescu, J. (2008, March 01). Research ethics and lessons from Hwanggate: What can we learn from the Korean cloning fraud? Retrieved from https://jme.bmj.com/content/34/3/214?legid=medethics;34/3/214

dealing with human subjects and research, clinical research protocols must adhere to government regulations. In the United States, this is the relevant law known as 45 CFR 46 subpart A, also termed the Common Rule; in Korea, Arts. 15-19 of the Bioethics and Biosafety Act. Both laws also cover specimens, data and identifiable information gathered about human subjects. Identification can come in the form of private identity information, coded specimens that are linked to the identity and can be traced back, or human tissue samples depending on the research. Voluntary informed consent is often a prerequisite for involving human subject in research. Sometimes, even if the researcher is not recruiting patients for studies, simply studying or analyzing previously collected patient data or specimens can be regulated as human subject research, which mandate IRB approval prior to the research. This is because research activities involving human subjects must receive IRB approval for proposed uses, both originally and subsequently. There are multiple examples of human subject research, such as administering tests on human specimens or identifiable data, have access to databases that identify subjects, have access to key-codes for specimens, and other items.

Seoul National University Research Ethics Guidelines

Revised on July 16, 2010.

Chapter 5 Ethics of Research on Human Subjects

Article 20

【Principle】

① A researcher of research on human subjects (including clinical test, biomedical science, behavioral science, etc.) shall uphold the human rights, health, and welfare of each human subject or research.

② The voluntary consent of a human subject of research shall be obtained after the purpose and procedure of research and expected risk, etc., have been sufficiently explained to a human subject of research. However, cases wherein prior permission from an official organization such as the Seoul National University Institutional Review Board (hereinafter referred to as "IRB") has been obtained may be excluded.

③ The risk to human subjects of research shall be minimized, and research that seriously affects the health and welfare of a human subject of research shall not be performed.

④ A human subject of research belonging to the vulnerable social group shall be specially taken care of in relation to the consent-obtaining process or risks.

⑤ The privacy of each human subject of research shall be secured, and his/her personal information shall be protected

20. IRB

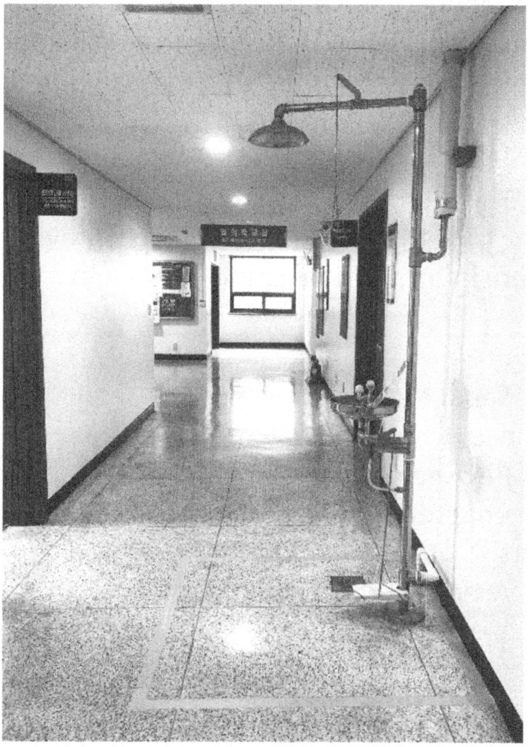

As mentioned, any human subject research must first be approved by the Institutional Review Board (IRB). Previously, use of an IRB was not as common, but thanks to growing awareness, it is now an integral part of the research ethics oversight process. Due to an emphasis on speed in research and the Korean emphasis on doing things quickly, ethical reviews by the IRB have sometimes been seen as a nuisance that hinders research progress.[169] As noted by Kim Jin-Soo, director of SNU's IBS Center for Genome Engineering, prior to intense media coverage, IRB oversight was ignored and procedures such as blood draws were routinely performed on a whim, but since then IRB approval has come to be recognized as necessary for any human subject experiments.[170]

The IRB has to monitor several aspects of a study or trial. Making sure all subjects provide voluntary informed consent, is just one part.

> Seoul National University Research Ethics Guidelines
>
> Revised on July 16, 2010.
>
> Chapter 5 Ethics of Research on Human Subjects
>
> Article 21
>
> 【Review by IRB】
>
> ① A human subject research project shall be commenced after the research proposal is submitted to and approved by the Seoul National University IRB.
>
> ② Even in the case of a research project

---

[169] Nho, H. (2016). Research ethics education in Korea for overcoming culture and value system differences. *Journal of Open Innovation: Technology, Market, and Complexity, 2*(1). doi:10.1186/s40852-016-0030-3

[170] Zastrow, M. (2016). Why South Korea is the world's biggest investor in research. *Nature, 534*(7605), 20-23. doi:10.1038/534020a

that involves faculty, staff, or students belonging to Seoul National University as a subject, the provision of Paragraph (1) shall apply.

③ The Seoul National University IRB shall make an independent deliberation and supervise the research process in accordance with an appropriate procedure.

21. Animal Experiments

Much like the IRB supervises research involving human participants in research, the Institutional Animal Care and Use Committee (IACUC) deals with research involving animal experiments. In the wake of Korea's Animal Protection Act (Law No. 10310) and the Laboratory Animal Act (Law No. 9932) the SNU's IACUC has helped to protect animals from unethical handling and ensure that university research involving animals meets certain guidelines. The IACUC is comprised of no more than 15 people appointed by the university president, and currently has roughly 13 members with 5 commonly belonging to non-university affiliations. The IACUC also has a chairperson elected among committee members. Together, these members of the IACUC discuss and analyze the ethical issues regarding approval for animal experiment plans and the actual management and disposal of laboratory animals. Any other issues, including facility maintenance for animal facilities, are handled by this.

Because animals are so often necessary for use in biomedical research, the IACUC is tasked with reviewing concerns involving their use in animal research. As research on animals is often a precedent for research of the same variety involving humans, the IACUC role's is both functional and pragmatic. At the same time, they are ethically focused on preventing excessive and egregious pain and suffering for animals. At other times, the IACUC may recommend alternatives to the use of animals in research, or a change in the type of animal suggested for experimentation.

Seoul National University Research Ethics Guidelines

Revised on July 16, 2010.

Chapter 6 Ethics of Research on Animal Subjects

Article 23

【Principle】

A researcher of research on animal subjects shall respect the health and welfare of laboratory animals and perform the research scientifically and ethically by complying with the principles in the following subparagraphs:

1. Principle of replacement: A non-animal

model or a lower form of animal, if possible, shall be used; 2. Principle of reduction: Various methods to reduce the population size of laboratory animals shall be utilized; and 3. Principle of refinement: The pain and anxiety of laboratory animals shall be removed or minimized.
Seoul National University Research Ethics Guidelines Revised on July 16, 2010. Chapter 6 Ethics of Research on Animal Subjects Article 24 【Approval by the Seoul National University Institutional Animal Care and Use Committee】 An animal subject research project shall be commenced after the research proposal is submitted to and approved by the Seoul National University Institutional Animal Care and Use Committee.
Seoul National University Research Ethics Guidelines Revised on July 16, 2010. Chapter 6 Ethics of Research on Animal Subjects Article 25 【Close of Research on Animal Subjects】

When research on animal subjects is closed, the following matters shall be observed: 1. An animal on which the experiment was completed shall be euthanized in a humane way; and 2. Carcasses of laboratory animals shall be stored as determined by the head of each laboratory animals facility and disposed of according to the Infectious Waste Disposal criteria under the Waste Control Act.
Seoul National University Institutional Animal Care and Use Committee Regulations Revised on May 4, 2015. Article 8 【Deliberation, etc.】 ① Any researcher intending to perform an animal test in SNU shall submit an application form for animal test to the Committee and commence the animal test only after obtaining approval. ② The Committee shall notify the animal test implementer of the deliberation result for the application form for animal test under Paragraph (1). ③ Documents relating to animal tests shall be preserved for three (3) years after the closure of the relevant research. ④ The Committee shall decide whether to approve an animal test considering the following matters: 1. Whether or not the welfare of the animal is taken into consideration;

2. Whether or not the method of minimizing the suffering and fear of the animal or its alternative method is appropriate;

3. Whether or not the reason for using the animal and the species and number of used animals are appropriate:

4. Whether or not the method of euthanasia of the animal is appropriate;

5. Whether or not there is any operation that is not necessary or is likely to aggravate the conditions of the animal;

6. Whether or not aseptic surgery is performed and veterinarian technique before and after surgery is used;

7. Whether or not an experimental facility suitable for the animal and which considers the researcher's safety is prepared;

8. Matters concerning the appropriateness of an animal test that uses a toxic substance [Newly established on January 17, 2001]; and

9. Other matters concerning the ethics and science of animal tests.

22. Collaborative Research

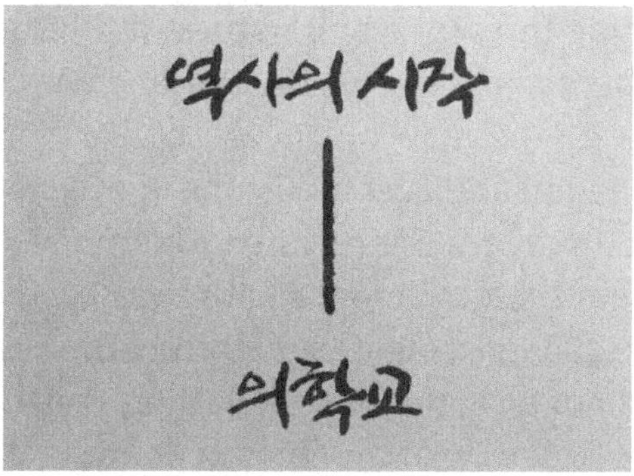

Collaboration as an organic link between different organizations, institutions, and researchers can bring new ideas and perspectives to research. Despite longstanding barriers and cultural constraints to collaboration, recently Korea has begun to foster more collaboration opportunities to connect researchers who may belong to different affiliations. Collaborations have positive benefits such as promoting the mutual exchange of ideas and innovation, but can also bring risks such as transfer agreements or additional IRB oversight. These risks need to be discussed before collaboration begins, and outline different ground rules for the work to be done. However, there are difficulties in implementing this practice due to the influence of cultural norms in Korea. Moreover, because leadership is often directed by Korean government goals, it can be difficult for collaborations to address these issues on their own from an independent point of view.

The implication is that sometimes in collaborations, especially international collaborations, it is better for collaborators to take the lead themselves in determining the scope and nature of their goals, commitments, and norms for working with others.

The joy and accomplishment of a new scientific discovery are rewards that many scientific researchers seek to achieve with their work. However, such discoveries are rarely the result of the efforts of a single person. Rather, collaboration can help reduce error and bias in analysis, expand scientific knowledge, enhance public trust, and strengthen interdisciplinary cooperation.[171] However, it is important to confront entrenched ways of thinking to joint cooperative research. At many public universities, academics view professional success in terms of tenure and promotions. These have typically been based on publication counts, as a zero-sum method of authorship: as the single author of a paper, they would theoretically earn more credit than as co-author on a four-author paper. Of course, this notion has been greatly discredited in many areas of science, and in fact, the number of authors per publication is actually growing in many fields. Still, with a system that rewards author isolation, international collaborations in particular that

---

[171] Nho, H. (2016). Research ethics education in Korea for overcoming culture and value system differences. *Journal of Open Innovation: Technology, Market, and Complexity, 2*(1). doi:10.1186/s40852-016-0030-3

support cutting-edge science may be harder to justify on a selfish level versus meaningless, assembly line science published in minor domestic journals with limited global exposure.[172]

Indeed, the previous experience of Korea with international collaborations is again reflected by the negative aftereffects of the stem-cell media coverage from 2004. The professor had achieved success within his field of animal science, but failed to garner attention with successful publications. To remedy this, he recruited an American co-author, working within the United States. However, the co-author was not required to use IRB oversight because no US federal funding was used. At the time, federal definitions did not mandate that type of research with no human subjects include IRB review since the samples had been removed of donor information that could link samples to donors.[173] This view on research involving human tissue samples has since been changed, with new US definitions of research implemented newly in 2018. Transnational research can be complex and involve loopholes in regulations or excessive regulatory constraints. However, Korean law covering organ sharing actually should have alerted

---

[172] Zastrow, M. (2016). Why South Korea is the world's biggest investor in research. *Nature, 534*(7605), 20-23. doi:10.1038/534020a

[173] Savulescu, J. (2008, March 01). Research ethics and lessons from Hwanggate: What can we learn from the Korean cloning fraud? Retrieved from https://jme.bmj.com/content/34/3/214?legid=medethics;34/3/214

both co-authors that US IRB review was required, because Korean law mandated that if therapies from research arise, they must benefit donors specifically.[174] Collaborators need to ensure that they work with individuals who are trustworthy, as lies about his research can cause a loss in international prestige, possibly forever.[175]

In addition to authorship, inventorship of patent and ownership of copyright and patent arising from research efforts can be disputed among collaborators. The inventorship is determined by patent law. Although an inventor is also a (co-)author in many cases, the criteria of inventorship is different from that of authorship. The copyright resides with the author, unless the work was commissioned as a work-for-hire. In fact, the authorship of a work-for-hire will also reside with the employer. In contrast, the ownership of a patent is determined by the employment agreement, IP ownership agreement, and hired-to-invent agreement, if there is any. If an employer's resources or time was used to invent, the employer can have a non-exclusive "shop right" to practice inventions owned by employee.

---

[174] Savulescu, J. (2008, March 01). Research ethics and lessons from Hwanggate: What can we learn from the Korean cloning fraud? Retrieved from https://jme.bmj.com/content/34/3/214?legid=medethics;34/3/214

[175] Cyranoski, D., & Check, E. (2005). Clone star admits lies over eggs. *Nature, 438*(7068), 536-536. doi:10.1038/438536a

At a public university such as Seoul National University, the ownership and control issues can be more complex. For example, a public university is usually using government funds and tuition monies to pay for the basic laboratory and its upkeep. The researchers receive a grant from a government agency to conduct specific research and/or a non-profit source for closely-related research. The laboratory and its staff are conducting other related work under funding through a sponsored research agreement with a private corporation. In such a situation, any patentable inventions that the researcher invents in the course and scope of her employment with the university must be assigned to the university. The university, not the funding agency, will only get title to an invention funded, even just in part, by governmental agencies, in the event that the university formally elects to take such title under the provisions of the Technology Transfer and Commercialization Promotion Act in Korea and the Bayh-Dole Act in the U.S.

Alternatively, non-profit funding sources in this situation may require assignment of any IP arising from the research, or may require certain uses or limitations on what the university can do with the IP, even where the university retains title to it. Private for-profit commercial firms rarely sponsor research without requiring assignment of any resultant IP, and further, generally institute strict confidentiality provisions on lab staff. Therefore, it is plausible that patentable inventions should be assigned to more than one funding source.

23. Mentoring

Mentors are those who help mentees, trainees, and beginning researchers in their career progress. They provide an introduction and guidance to the world of research and help navigate the unwritten rules of behavior. Good mentors allow those who they mentor to succeed past themselves, reaching even greater heights eventually than even the mentor does. A poor mentor, in contrast, believes mentees should remain subservient and below them until the mentor retires.

Mentorship can also be important to promote equity and fairness to prevent gender-based discrimination. Mentors, including female professors, can guide female researchers and promote better representation for them among faculty and leadership roles.

Good mentorship involves the respectful and fair treatment of others. In 2018, a leading genetic researcher at an English cancer research institute lost a $4.5 million grant due to allegations of systemic bullying within her laboratory.[176] In Korean society, leader-centric decision-making can result in laboratory and research environments that do not reflect consensus of lower-level workers, instead focusing on dominant judgment and self-dealing by professors and seniors.[177]

Finding a good mentor can be a challenge for beginning researchers. Researchers in need of mentoring should research past publications and professional expertise of potential mentors, as well as reviewing new developments in areas that professors may not have written about since then. Thoughtful questions and suggestions can spark a connection, and actually visiting potential mentors and laboratories, so these tools can be a useful way of assessing and how mentors treat others and will serve as a mentor.

---

[176] Else, H. (2018). Top geneticist loses £3.5-million grant in first test of landmark bullying policy. *Nature, 560*(7719), 420-420. doi:10.1038/d41586-018-06009-9

[177] Nho, H. (2016). Research ethics education in Korea for overcoming culture and value system differences. *Journal of Open Innovation: Technology, Market, and Complexity, 2*(1). doi:10.1186/s40852-016-0030-3

24. Conclusion

The nature of scientific research involves the incorporation of past results with current findings in order to build upon the body of work that comprises the aggregate knowledge of human effort and spirit. In essence, new research is not viewed in isolation without regard to contemporaneity, chronology, or applicability; it is an integral part of the overall scientific community and academic field.

Korea continues to invest more and more funding into research and development. In 2014, over ₩63.7 trillion was invested in research and development, making up roughly 4.29% of domestic GDP. This figure is far ahead of similar outlays as a percentage of GDP by other nations, and classifies Korea as potentially the largest investor in research in the entire world. From 2011-2015, almost 100,000 papers were published cumulatively in the areas of medical and health sciences by Korea, more than any other discipline.[178] As such, clinical research ethics has a leadership position to help other disciplines domestically as well as internationally.[179] While much of the focus of research is geared towards attaining prizes or awards such as Nobel Prizes and achieving success similar to regional rivals such as Japan or international leaders such as the United States, it is important to remember that Korea has already achieved international recognition in many areas of science. It is critical that Korea acknowledge its own leadership role and serve as a model for ethical research for other Asian nations and universities abroad. Moreover, Korea enjoys international collaboration opportunities in research. According to the United States National Science Foundation, almost 70% of Koreans who received PhD degrees from the United States during the period of 2008 to 2011 planned to remain in the United States.[180] Korean researchers have also been involved in studies in Southeast

---

[178] Zastrow, M. (2016). Why South Korea is the world's biggest investor in research. *Nature, 534*(7605), 20-23. doi:10.1038/534020a

[179] Zastrow, M. (2016). Why South Korea is the world's biggest investor in research. *Nature, 534*(7605), 20-23. doi:10.1038/534020a

[180] Zastrow, M. (2016). Why South Korea is the world's biggest investor in research. *Nature, 534*(7605), 20-23. doi:10.1038/534020a

Asia and Northeast Asia contexts, with nationals from these countries also making up a growing percentage of Korean students and researchers. As such, this textbook attempts to invoke ethical principles for international topics given the increasingly borderless nature of research.

New researchers who have just begun to be exposed to ethical principles in research can definitely benefit from learning about ethics in research. However, mere understanding is not sufficient. Mandatory, required ethics training for every student can change the perspective on ethical conduct before students are exposed to precarious situations later in their careers. Developing the personal tools and skillsets to mitigate risks and incentives for misconduct are useful for any researcher and scientist.

* * *

Credits

Biomedical Research Ethics
by Mikyung Kim and Christopher B. Shook

Mikyung Kim, MD, PhD, JD
Mikyung Kim is a Professor in the Department of Legal Medicine, College of Medicine, at Seoul National University, Seoul, Republic of Korea.

Christopher B. Shook, JD, MBA
Christopher Shook is a Visiting Assistant Professor in the Department of Legal Medicine, College of Medicine, at Seoul National University, Seoul, Republic of Korea.

Made in the USA
Monee, IL
04 November 2026

67382584R00089